A Primer in
Data Reduction

A Primer in Data Reduction

An Introductory Statistics Textbook

A. S. C. EHRENBERG
London Business School

JOHN WILEY & SONS

Chichester · New York · Brisbane · Toronto · Singapore

Library of Congress Cataloging in Publication Data:

Ehrenberg, A. S. C.
  A primer in data reduction.

  Includes index.
  1. Statistics.   2. Data reduction.   I. Title.
QA276.12.E37   001.4′22   81-16360

ISBN 0 471 10134 6 (Cloth)   AACR2
ISBN 0 471 10135 4 (Paper)

British Library Cataloguing in Publication Data:

Ehrenberg, A.S.C.
  A primer in data reduction.
  1. Statistics
  I. Title
001.4′22   HA29

  ISBN 0 471 10134 6 (Cloth)
  ISBN 0 471 10135 4 (Paper)

Typeset by Pintail Studios Ltd, Ringwood, Hants
Printed in Great Britain at The Bath Press, Avon

To H.B.L. and C.L.E.

Contents

Preface

This is a basic text in statistics. It is for non-specialists taking an introductory course in the subject.

Faced with increasing amounts of numerical data as part of the wider information explosion, we need to know how to reduce such data to statistical summaries and how to interpret the results. This book helps the reader develop these skills, using practical and varied examples.

It differs from most other introductory texts in two ways. First, it covers a wider range of relevant issues. In particular, it not only describes traditional statistical methods but also notes their limitations and discusses alternatives. Secondly, students have found it easier to understand: it avoids most mathematics but without oversimplifying the underlying concepts, it explains more, and it is briefer.

The book is in six parts:

Part One deals with averages and measures of scatter.

Part Two covers frequency distributions and probability.

Part Three is on sampling: How sample data are collected; how errors due to sampling can be measured; and how statistical inferences are made.

Part Four discusses relationships between observed variables. It covers standard techniques (such as correlation, regression, and factor analysis) and also the procedures which lead to the law-like relationships of science.

Part Five is about communicating results. It shows how to make numbers, tables, and graphs easier to grasp. It also gives guidelines for writing and presenting technical reports.

Part Six puts statistics into the wider context of the empirical generalizations and laws of science. It discusses the nature of description and explanation, and the design and interpretation of experiments and observational studies.

The later sections in each chapter may be treated as optional or skimmed at a first reading. Each chapter closes with some exercises. Most have a glossary of technical terms. Answers to the exercises and some statistical tables are given at the end of the book, together with a comprehensive subject index and a short list of symbols and formulae.

We often have to use numbers in our studies and later working-lives—to describe and interpret what has happened, to forecast what will happen, to analyse trends and developments, to explain, and to plan. The book aims to help readers to understand and communicate numerical data, and to judge how statistical methods work and how far they make sense. There has never been a greater need for this kind of numeracy.

London Business School   A. S. C. EHRENBERG
January 1982

Teachers' Preface

Students are often put off by statistics and statistics courses. They claim they feel 'less numerate at the end of the course than at the beginning' and that 'most of the course seems irrelevant to what we are doing.' The students are usually right. Traditional statistics courses *are* largely irrelevant—not just boring or technically difficult, but irrelevant to their other work.

Students who are not specializing in statistics do not need to master techniques which they will not subsequently meet or use. For example, they will seldom need to test the statistical significance of an isolated result. That is not the kind of empirical evidence which they will often meet in their main courses, at least not without ample warning from their teachers that it is an isolated one-off result which has never been repeated. Nor will most of them be doing much original research based on random samples, at least not without some further statistical training.

The book therefore plays down (but still covers) statistical inference from samples and tests of significance. It brings out, as an addition, the central role of empirical generalizations, showing them as results which are known to hold under a wide range of different conditions and which therefore provide the traditional basis for scientific prediction.

The book generally concerns itself more with data than with techniques or with proofs which do not aid the student's understanding. In avoiding unnecessary mathematics, it even says 'Sum' instead of Σ. (Some teachers may find 'Sum' uncomfortable, but Σ is much worse for many students at the start of an introductory course.) The book also faces up to the fact that some statistical techniques are less widely used, or have led to fewer useful results, than one might gather from the traditional texts.

Structure

The Primer derives from an approach developed in my earlier *Data Reduction* (Wiley, 1975, 1978). But the material is now geared to the needs of a one-semester introductory course.

The first fourteen chapters contain the material common to standard syllabuses. Interspersed is additional material on looking at tables of data (Chapter 3) and on the law-like relationships of science (Chapter 13). A teacher pressed for time can leave out some or all of Chapters 3, 6, 8, 13, and 14, as well as the later parts of Chapters 5 and 10, without harming the basic structure of the course. Parts Five and Six can be treated as further reading for the student.

The book gives more coverage to some topics than basic texts normally do. For example, theoretical frequency distributions are treated in Chapters 5 and 6 as an instance of model-building. There is also more emphasis on relationships, since they are central to the analysis of observed phenomena. Chapter 14 faces up to multiple regression and factor analysis, as many students will come across one or other of these techniques in their main studies. Other material contained here which is seldom covered in introductory texts is on communicating results (Part Five) and general scientific method (Part Six).

The book basically is intended to suit introductory statistics courses given for students in the social, biological, and natural sciences, in management, and in professional subjects. The illustrative examples used in the text are varied and realistic.

Specific subject-areas will require added emphasis on particular topics or techniques. For instance, psychology and sociology students may want more examples on correlation and experimental design; medical students more on clinical trials and epidemiology; economics or business students more on index numbers, regression methods, time-series, sample surveys or decision theory; engineers more on quality control. Teachers of such classes will be able to add specialized material and illustrations and to advise their students on relevant supplementary reading.

There has long been a real need for a statistical text which successfully communicates basic statistics to non-specialists in a way which students see as being relevant to their other studies and to the world in which they will be working and living. Extensive class-room tests and reactions of teachers and students have shown that the Primer can do this better than most.

A teachers' guide is available from the publishers (John Wiley and Sons, 606 Third Avenue, New York, NY 10016, USA, or Baffins Lane, Chichester, Sussex, England) or from the author.

Acknowledgements

Successive drafts of this book have been used in courses at the London Business School and I am much indebted to numerous colleagues and students for their comments. More specifically, I wish to thank Mr Patrick Barwise, Mrs Myra Chapman, Dr Elroy Dimson, Professor Gerald Goodhardt, Dr Stuart Hodges, Ms L. Ann Law, Mrs Mary Jackson, Dr Spyros Lioukas, Mr James M. Maloney,

Professor Harry Smith Jr, Mr William Templeton, and Mr Daniel Webster. The book's clarity has however been aided most of all by the hard work and constructive insights of my critic and friend, Helen Bloom Lewis, to whom the reader and I have to be exceptionally grateful.

London Business School   A. S. C. EHRENBERG
January 1982

Detailed Contents

PART ONE: STATISTICAL DATA

We are often faced with variable measurements like daily temperatures, people's heights, crop yields, etc. A basic task in statistics is to reduce such data to a brief summary. We usually do so by working out averages and summarizing the variation about these averages. This is discussed in Chapters 1 and 2.

Much data come in the form of tables where the figures are arranged in rows and columns. The analysis of tables is introduced in Chapter 3.

CHAPTER 1

Averages

Averages are the main tool in analyzing statistical data. We tend to use them almost automatically when describing a group of numbers. How old are the students in the class? The average age is about 21.

Such a summary works well when most of the individual measurements are close to the average. But if the readings vary greatly, the average may take an atypical value and could be misleading. In such cases a fuller description of the distribution of readings must also be given.

1.1 The Arithmetic Mean

The arithmetic mean, or mean for short, is the most important type of average. It is defined as the sum of the given set of readings divided by the number of readings:

$$\text{Mean} = \frac{\text{Sum of readings}}{\text{Number of readings}} \quad \text{or} \quad \frac{\text{Sum of readings}}{n}.$$

The symbol n denotes the number of readings.

As an example, Table 1.1 gives the test grades of 20 students. The sum of the readings is 108. Dividing this by $n = 20$ gives a mean of 5.4.

TABLE 1.1   Grades of Twenty Students

6, 3, 7, 5, 6, 4, 4, 6, 7, 3, 5, 9, 6, 4, 2, 7, 5, 5, 8, 6
Mean = 5.4

The pattern in the data can be seen more easily if the readings are arranged in order of size, as in Table 1.2.

TABLE 1.2 The Twenty Grades in Order of Size

| 2, | 3, 3, | 4, 4, 4, | 5, 5, 5, 5, | 6, 6, 6, 6, 6, | 7, 7, 7, | 8, | 9 |

Mean = 5.4

We can see now that the mean is a good summary figure. About half of the readings are smaller than the mean and half larger. While the readings range from 2 to 9, most of them lie closer to the mean.

Hump-Backed Symmetrical Distributions

By counting the number of times each value occurs in a set of readings we get a *frequency distribution*. For example, in Table 1.2 there is one 2, two 3s, three 4s, and so on.

 The shape of the frequency distribution determines whether the mean serves as a good summary figure for the data. We can see the shape more easily if we plot the frequencies of the readings on a graph, as in Figure 1.1. The distribution is almost symmetrical: about half of the readings are below the mean and half above it, with most of the numbers in the middle, where there is a peak or 'hump'.

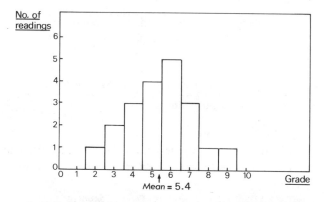

FIGURE 1.1 The Twenty Grades from Table 1.2

 Distributions which are hump-backed and roughly symmetrical like this are the only ones which are simple to summarize. With such data the mean gives a fairly close description of most of the readings.

 If the arithmetic mean is quoted without qualification, it should imply that the data follow a hump-backed, symmetrical distribution. In all other cases more detail should be given or implied by the context.

Skew Distributions

Table 1.3 gives an example of data with a non-symmetrical or skew distribution.

TABLE 1.3   The Number of Employees in Twenty Grocery Stores

2,	3, 3, 3,	4, 4, 4, 4,	5, 5, 5, 5, 5, 5,	6, 6, 6,	8,	10,	15

Mean = 5.4

The readings show the number of employees in twenty grocery stores: one store had two employees, three stores had three employees, and so on. The sum of the readings is 108 with $n = 20$, so the mean is again 5.4. But is this a good summary figure for these data?

Figure 1.2 shows the shape of the distribution graphically. It has a hump, but not in the middle. The distribution is not symmetrical: the tail to the right is longer than the tail to the left. While most of the readings are close to the mean, there are more readings just below 5.4 than just above it. This would have to be noted in any good summary of the data.

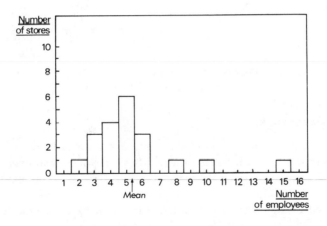

FIGURE 1.2   The Numbers of Employees in Grocery Stores

Table 1.4 gives a more extreme example of a skew distribution: the number of days of illness for a group of twenty people. Again the mean is 5.4. But one can see at a glance that this figure is not typical of most of the readings. It would be misleading to quote the mean as if most of the readings were close to 5.4 and distributed symmetrically about it.

TABLE 1.4   The Number of Days of Illness for Twenty People

0, 0, 0, 0, 0, 0, 0, 0, 1, 1, 1, 1, 1, 2, 2, 2, 3, 3, 15, 76	
Mean = 5.4	

A good summary of these data would say that most of the readings were 0, 1, or 2, with a few much higher ones. A distribution like this, with many low readings and a long tail to the right of a few higher ones, is usually called reverse-J-shaped. The name is explained by the shape in Figure 1.3.

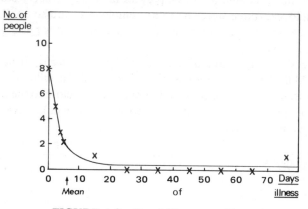

FIGURE 1.3   Days' Illness per Year
(Observed frequencies and free-hand curve)

U-Shaped Distributions

Sometimes a distribution has more than one peak. Table 1.5 gives an example with data on the number of issues of a monthly magazine read over a year by a group of twenty people. Again the mean is 5.4 and again it is not a typical figure.

TABLE 1.5   The Number of Issues of the Monthly Magazine X
Read per Year by Twenty People

0, 0, 0, 0, 0, 0, 0, 0, 1, 1, 1, 2, 10, 11, 11, 11, 12, 12, 12, 12, 12	
Mean = 5.4	

Most of the readings are well away from the mean. No one read five or six issues, or anything like that. This kind of distribution, with peaks at each end, is called U-shaped. It is shown in Figure 1.4.

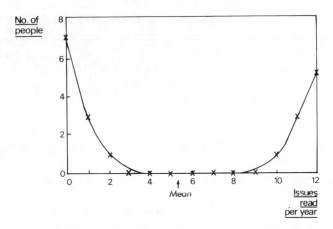

FIGURE 1.4   Number of Issues Read per Year
(Observed frequencies and free-hand curve)

Even though the distribution is fairly symmetrical, there is no typical reading here. Thus any summary of these data would have to be more complex than quoting the mean. One would have to say that just over half the people read none or almost none of the issues, and the other half read all or nearly all.

Still another case is the *Uniform* distribution, where roughly equal numbers of readings occur at each possible value. If we look at the distribution of ages of people in their thirties, we might find roughly 10 per cent aged 30, 31, 32, etc. Uniform distributions are symmetrical but do not have a hump or peak anywhere.

Broadly speaking then there are two kinds of data:
(1) Roughly symmetrical distributions with a peak or hump in the middle, where the mean is a good summary figure because it is fairly typical of most of the readings.
(2) Skew, U-shaped, Uniform, or more complex distributions, where there is usually no typical reading and the mean by itself is not a good summary figure.

1.2   The Mean as a Focus

Even when the mean is not a good summary figure, it generally provides a useful mental or visual focus when looking at the data. So it is a good idea to calculate the mean of a data set in the early stages of analysis. At the very least this can make it easier to see whether the data are symmetrical or skew.

To illustrate, Table 1.6 reproduces the four earlier sets of readings, but not yet ordered by size.

It is not easy to see any patterns in the data. But if we know that the mean of

TABLE 1.6 Four Sets of Twenty Readings

Table 1.1 :	6, 3, 7, 5, 6, 4, 4, 6, 7, 3, 5, 9, 6, 4, 2, 7, 5, 5, 8, 6
" 1.2 :	4, 6, 5, 3, 15, 4, 5, 5, 5, 2, 8, 4, 3, 10, 5, 6, 5, 6, 3, 4
" 1.4 :	2, 0, 1, 0, 0, 3, 1, 0, 76, 2, 0, 2, 1, 1, 3, 1, 0, 0, 15, 0
" 1.5 :	11, 0, 12, 1, 0, 12, 1, 0, 0, 12, 12, 2, 11, 10, 0, 12, 1, 0, 0, 11

each set is about 5, we can look along each line of readings with this single figure in mind. We can then see that

- In the first line most of the readings are within 1 or 2 units of 5, with about half higher and half lower than the mean.
- In the second line most of the readings are about 5 or a little less, with only two readings much higher.
- In the third line nearly all of the readings are well below 5, with two high readings of 76 and 15 standing out.
- In the last line almost half of the readings are well above 5 (mostly at 11 or 12) and the others are well below 5 (at 0 or 1).

This focusing function of the mean applies even when the readings have already been ordered by size, as in Table 1.7.

TABLE 1.7 The Four Sets of Readings, Ordered by Size

2, 3, 3, 4, 4, 4, 5, 5, 5, 5, 6, 6, 6, 6, 6, 7, 7, 7, 8, 9	Mean: 5.4
2, 3, 3, 3, 4, 4, 4, 4, 5, 5, 5, 5, 5, 5, 6, 6, 6, 8, 10, 15	Mean: 5.4
0, 0, 0, 0, 0, 0, 0, 0, 1, 1, 1, 1, 1, 2, 2, 2, 3, 3, 15, 76	Mean: 5.4
0, 0, 0, 0, 0, 0, 0, 1, 1, 1, 2, 10, 11, 11, 11, 12, 12, 12, 12, 12	Mean: 5.4

With data ordered by size one can also mark the mean physically, as in Table 1.7a, which helps visually.

TABLE 1.7a The Means Drawn in by Hand

2, 3, 3, 4, 4, 4, 5, 5, 5, 5, 6, 6, 6, 6, 6, 7, 7, 7, 8, 9	Mean: 5.4
2, 3, 3, 3, 4, 4, 4, 4, 5, 5, 5, 5, 5, 5, 6, 6, 6, 8, 10, 15	Mean: 5.4
0, 0, 0, 0, 0, 0, 0, 0, 1, 1, 1, 1, 1, 2, 2, 2, 3, 3, 15, 76	Mean: 5.4
0, 0, 0, 0, 0, 0, 0, 1, 1, 1, 2, 10, 11, 11, 11, 12, 12, 12, 12, 12	Mean: 5.4

1.3 Comparing Different Sets of Data

The analysis and interpretation of statistical data usually involves comparisons of different sets of readings. Are boys taller than girls? Did a treated group of patients fare better than an untreated group? Instead of comparing all of the individual readings in the different sets, we can compare the means—as long as the sets being compared have the same distribution shape.

Table 1.8 sets out the number of headaches suffered in a month by sixteen treated and twenty untreated patients.

TABLE 1.8 Comparing Two Sets of Readings

Treated:	1, 2, 2, 3, 4, 4, 5, 5, 5, 6, 6, 6, 7, 7, 8, 9	Mean: 5.0
Untreated:	4, 5, 6, 7, 7, 7, 8, 8, 8, 8, 8, 8, 8, 9, 9, 9, 9, 10, 10, 12	Mean: 8.0

The two groups have different numbers of patients and different means, 5 and 8, but they are otherwise similar: the individual readings are spread in the same way, within 3 or 4 units on either side of the means. Thus we can summarize the data by saying that the untreated patients had on average three headaches more that month than the treated patients.

Both data sets here had hump-backed and roughly symmetrical distributions, so the means were good summary figures. But the means can be used for comparison even with skew or U-shaped distributions, as long as their shape is the same.

Table 1.9 gives an example for skew distributions. The means of 12.4 and 5.4 are not typical of the readings, but they still bring out the main difference between the sets: many of the readings in Set X are substantially bigger than those in Set Y.

TABLE 1.9 Two Skew Distributions and their Means

Set X:	1, 1, 1, 2, 3, 5, 8, 12, 23, 68	Mean: 12.4
Set Y:	1, 1, 1, 1, 2, 2, 3, 4, 6, 33	Mean: 5.4

1.4 Medians and Modes

Two other measures which are sometimes used to describe the average level of a set of readings are the median and the mode.

The *median* is defined as the point in the middle: 50 percent of the readings lie below it and the other 50 percent lie above it. This definition does not always apply exactly, but we can make common-sense adjustments. For example, for Set X in Table 1.9 we take the median to be at 4, halfway between the 3 and the 5. For Set Y the median is taken to be at 2, since 50 percent of the readings are 2 or less and 50 percent are 2 or more.

The *mode* is the most frequent (or fashionable) reading in a data set. For the reverse-J-shaped data of Table 1.9, the modes are at 1. With the symmetrical hump-backed distributions in Table 1.8, the modes are at 5 for the treated patients and at 8 for the untreated patients.

Sometimes data have more than one mode. A set like 1, 4, 5, 5, 6, 6, 7, 8 has equal numbers of 5s and 6s and hence *two* modes. With a U-shaped distribution, like the one in Figure 1.4, it is possible to speak of two 'local' modes at 0 and 12, even though the frequency of 12 is smaller. (A 'local mode' is defined as a value that is more frequent than the other values near it.)

When the frequencies in a data set are all small, it is pointless to speak about a most frequent or modal value. For example, with the age distribution in Table 1.10 it makes no sense to pinpoint 36 as the mode.

TABLE 1.10   The Ages of Twenty People

24, 25, 28, 29, 31, 32, 34, 35, 36, 36, 37, 38, 39, 42, 44, 45, 48, 51, 53, 55

But ages in the thirties are the most common in the data. They show up as the most frequent or modal category when the readings are grouped in 10-year intervals, as in Table 1.11. Here a single modal value might be taken as 35. But

TABLE 1.11   The Age-Distribution Grouped

	20-29	30-39	40-49	50-54
Number	4	9	4	3

with 5-year groupings the 35–39 category would be the most frequent, with a single modal value at 37.5 or so. This shows how the precise value of a mode can depend on the grouping or units of measurement used. The mean and the median are affected less.

Comparing the Mean, Median, and Mode

The three types of average measure should not be confused. They describe three quite different things.

The arithmetic mean: the sum of the readings divided by *n*.

The median: the half-way point in a set.

The mode: the most frequent reading in a set.

With roughly symmetrical, hump-backed distributions the values of the mean, median, and mode are about equal. That is why such distributions are easy to describe. But with other kind of distributions the three measures usually take different values. Table 1.12 shows this for our earlier examples.

TABLE 1.12   The Mean, Median, and Mode for Three Distribution Shapes

		Mean	Median	Mode
Approx. Sym. humpb.	(T.1.2):	5.4	5.5	6
Reverse J-shaped	(T.1.4):	5.4	1.0	0
U-shaped	(T.1.5):	5.4	1.5	0 & 12

When the three summary figures are unequal it is a sign that the distribution is markedly non-symmetrical and/or not hump-backed. The extreme case was the U-shaped distribution: the mean was 5.4, the median was 1.5, and there were two modes, at 0 and 12.

The Pre-eminence of the Mean

The median and mode can provide us with a general feel for the data by showing where the 50 percent point comes and what the most frequent reading is. But neither measure is useful in detailed analysis, nor usually as the main summary figure when reporting data.

The arithmetic mean has three properties which tend to make it the most useful average measure:

(i) The mean can be calculated by summing the readings in any order. In contrast, to work out the median or mode the readings first have to be arranged in order of size. This can be very bothersome when there are large numbers of readings, even if the data are on a computer.

(ii) If the mean and the number of a set of readings are reported, one can easily recalculate the sum total of the readings, which is sometimes useful. (The mean = Sum of readings/n, so that $n \times$ mean = Sum of readings.) There is no corresponding calculation for the median or mode.

(iii) More generally, the mean as a summary figure can be used in subsequent analyses of the data, while the median or mode seldom can be. An example is when two or more data sets are to be combined.

To illustrate the latter point, Table 1.13 gives two sets of four readings. From the

TABLE 1.13   Two Sets of $n = 4$ Readings

		Mean	Median	Mode
1st Set	2, 3, 3, 4	3	3	3
2nd Set	3, 5, 5, 7	5	5	5
Both Sets	2, 3, 3, 3, 4, 5, 5, 7	4*	3.5	3

$$* \; \{(4 \times 3) + (4 \times 5)\}/8$$

two means, 3 and 5, we can calculate the mean of the combined group, 4. But we cannot calculate the median or mode of the combined group unless we return to all of the individual readings. At best this can be very cumbersome to do. In practice it is usually impossible, since the individual readings are rarely reported. (The whole point of statistical analysis is to summarize.)

The mode has another special drawback: different J-shaped (or U-shaped) distributions can all have their mode or modes at the same points. The readings in one reverse-J-shaped distribution may tend to be larger than in another, but both modes may be at zero so they will not show this difference. In such cases the mode only tells us about the *shape* of the distribution and not about the average value of the readings.

Sometimes usage of the arithmetic mean is criticized because for very skew data it is atypical and is neither the most common reading nor the 50:50 value. But such criticism misses the point. It is not that the mean is an atypical figure for such data, but that for very skew data there is no typical figure.

1.5   Outliers

The description of data is complicated by any readings which lie way outside the bulk of the readings. Such exceptional readings are usually called 'outliers'.

We saw an example earlier in the illness data which are shown again in Table 1.14. Most of the twenty people were either not ill at all or ill for only 2 or 3 days. But one person was ill for 76 days. When such an outlier occurs, one must query whether it belongs to the data. Was it just a computing, clerical or measurement error? (Twyman's Law says that any figure that looks different or interesting is usually wrong.)

TABLE 1.14   The Number of Days of Illness of Twenty People

0, 0, 0, 0, 0, 0, 0, 0,   1, 1, 1, 1, 1,   2, 2, 2,   3, 3,   15,   76
Mean: 5.4

Even when an outlier appears to be reliable, it still makes the data difficult to describe because it can greatly affect the mean. Without the exceptional reading of 76, the mean for the illness data would be 1.7 instead of 5.4. Thus for a summary, it is often best to omit the outlier from the main calculations and report it separately.

Neither the median nor the mode are affected by outliers. For the illness data the mode is at 0 whether we count the outlier or not. Similarly, the median remains at 1. Yet the fact that these measures are unaffected by outliers can be a mixed blessing. If outliers occur, one needs to know about them. They cannot simply be ignored.

In most analyses one is dealing with many different data sets, and often each contains large numbers of readings (not like the small and simple data sets in the examples here). In such cases one looks mainly at the data summaries; often nothing else is reported. An unusual value of the mean then acts as a warning signal that there may be an outlier or two, so the detailed readings should be checked. The sensitivity of the mean to outliers is therefore quite an advantage.

1.6 Weighted Averages

When two or more data sets are combined, there are two ways of calculating a mean for the enlarged set: (i) taking the arithmetic mean of the separate means; (ii) taking the arithmetic mean of all of the individual readings. The two results generally differ somewhat, but often not by much.

Table 1.15 shows two small data sets with $n = 8$ and $n = 3$ readings and means of 2 and 4. The average of the two means is 3. But the average of all eleven readings is $28/11 = 2.55$.

TABLE 1.15   The Mean of Two Sets of Readings

		n	Sum	Mean
1st Set	0, 1, 1, 2, 2, 3, 3, 4	8	16	2
2nd Set	3, 4, 5	3	12	4
Both Sets	0, 1, 1, 2, 2, 3, 3, 3, 4, 4, 5	11	28	2.55

The mean of all of the individual readings in the combined set is usually called the *weighted mean* or *weighted average* because of a short-cut method used to calculate it. (This avoids going back to all of the original readings and is especially useful because the individual readings are often not reported.)

The short-cut involves first recalculating the sum of each data set by multiplying its mean by n. Thus in our example,

Sum of first set: $8 \times 2 = 16,$
Sum of second set: $3 \times 4 = 12.$

The two sums are then added and the sum total is divided by the combined number of readings, here 11:

$$\frac{(16 + 12)}{(8 + 3)} = \frac{28}{11} = 2.55.$$

The weighted average got its name because in this calculation the means of the data sets, 2 and 4, are 'weighted' by the proportion of the total readings they represent, 8/11 and 3/11:

$$\left(\frac{8}{11} \times 2\right) + \left(\frac{3}{11} \times 4\right) = 2.55.$$

A weighted average is generally more laborious to calculate than the straight average of the separate means, but it correctly reflects the number of readings in each set. This presents a certain dilemma. Which figure should be used?

If the straight average and the weighted average are about equal, it does not matter much which one is used. The easier straight average is then usually preferred. But if the two figures differ substantially, neither may represent a typical reading or be a good summary of the combined data. This situation arises when the means of the separate data sets differ markedly and the numbers of readings in each also differ. It is then often best to describe such data sets separately and use *neither* form of combined averages.

When one calculates the overall average only as a working-tool, without expecting it to be typical of the data as a whole, one can use the easier straight average.

1.7   Algebraic Notation

Algebraic symbols are a form of short-hand.

A measured quantity or variable is often denoted by the symbol x. This represents any value (possibly within a certain range). But in any particular case it can also take a specific value. For example, we can say that x represents the amount of rainfall on any day, and also that yesterday's rainfall was $x = .3$ inches. The crucial point is that a symbol like x can stand for both *any* value and some *specific* value, depending on the context. This makes the symbolism particularly useful.

'Memorable' symbols like h for height or r for rainfall are often better than the ubiquitous x. A special symbol, n, is generally used for the number of readings in a set of data.

Several symbols tend to be used to denote the arithmetic mean, just as different words are sometimes used, like average or mean. One symbol is m ('m for mean'). Another is $\bar{x}$ (termed 'x bar'); this is used when the individual readings are denoted by the variable x.

When two or more sets of readings are being analyzed, each set may be denoted by a different letter, like x and y. The means of the sets can then be denoted by $\bar{x}$ and $\bar{y}$. If the symbol m is used for the mean, a suffix is added to identify the mean of each set. Thus m_x stands for the mean of the x readings, and m_y for the mean of the y readings.

Another notation, mainly seen in more advanced work, uses x_i as a symbol for the variable x. Here the suffix i can take any value from 1 to n. Hence x_1 stands for the first reading in the data set, x_2 for the second reading, x_n for the last or nth reading and x_i for the general or 'ith' reading (where i can again take either *any* value from 1 to n or a specific value like $i = 2$ for x_2). The mean of the readings may still be denoted by $\bar{x}$ or sometimes by $x_.$ (termed 'x dot').

The operation of summing a set of readings is often represented by the symbol Σ, the Greek capital S, pronounced 'Sigma'. Thus Σx means 'Sum of the readings x'. Hence we can write the arithmetic mean of n readings of the variable x as Mean $= (\Sigma x)/n$.

The range of values summed is sometimes denoted in the statistical literature by various forms of subscripts and superscripts to the Σ. For example, the following expressions all denote the summation of n values of the variable x:

$$\sum_{i=1}^{n} x_i, \quad \Sigma x_i, \quad \overset{n}{\Sigma} x, \quad \Sigma x.$$

The first expression is the most explicit, meaning the sum of all x_i from x_1 for $i = 1$ to x_n for $i = n$. But it is also the most complex. In this text summation is generally denoted simply as 'Sum(x)' or 'Sum x'. The parentheses are used when they help to clarify the arithmetical process.

1.8 Discussion

The numerical examples in this chapter have generally been small. But standard ways of summarizing data like the arithmetic mean are particularly helpful when dealing with larger sets of readings as in Table 1.16, or when comparing many different sets of data.

TABLE 1.16 **A Larger Set of Readings**

36, 49, 171, 40, 38, 95, 62, 37, 375, 370 74, 62, 392, 29, 45, 164, 311, 206, 419, 5	$\bar{x}$ = 149

The most widely used summary measure is the arithmetic mean. It is relatively easy to calculate and can be useful in the initial analysis, in reporting the results, and in subsequent analyses. The median (the mid-point) and the mode (the most frequent reading) are mostly helpful in the initial stages, to give one a grasp and feeling for the data.

The distribution of readings in a data set influences the way it can be summarized. Hump-backed symmetrical distributions are the easiest to summarize because the mean, median, and mode are usually about equal. Skew, J-shaped, and multimodal distributions tend to be more difficult to summarize. They usually have no typical reading and may therefore require tailor-made descriptions.

(Sometimes a mathematical formula can be used, as is discussed in Chapter 5.)

The basic task in statistics is to reduce numerical information to a brief summary. But just giving the average of a data set is not enough. The scatter or variation of the readings about the mean must also be described. This is discussed in Chapter 2.

CHAPTER 1   GLOSSARY

Arithmetic mean

The sum of the readings divided by the number of readings or Sum$(x)/n$. Generally called the *mean* or *average* for short, and denoted by $\bar{x}$ or m.

Average

Strictly speaking, any measure of the general size of a set of readings. But the term usually refers to the arithmetic mean.

Frequency distribution

(Or *distribution* for short.) The number of readings at each value in the data.

Hump-backed distribution

A distribution with a strong mode or peak roughly in the middle.

Local mode

A value that occurs more often than the other values near it.

Mean

The arithmetic mean.

Median

The 50:50 point in a set of readings ordered by size.

Mode

The most frequent reading in a data set.

m

A common symbol for the mean.

n

The common symbol for the number of readings in a data set.

Outlier

An unusual reading much above or below the other values in a data set.

Σ (Sigma)

The capital Greek S. Usually stands for the notion of summing a set of readings.

Skew distribution

A non-symmetrical frequency distribution.

Symmetrical distribution

A frequency distribution where the left half is the mirror image of the right half.

Unweighted average

The average of the means of two or more data sets.

Variable

A measured item which can take different numerical values, often denoted by x.

Weighted average

The average of the means of two or more data sets, where each mean is multiplied or 'weighted' by the proportion of the total number of readings it represents. (It equals the average of all of the individual readings in the combined sets of data.)

x

A general symbol used to denote a variable.

$\bar{x}$

Pronounced '*x* bar'. A symbol for the mean of a set of readings denoted by x.

y

Another general symbol used to denote a variable when it is to be differentiated from variable x.

CHAPTER 1   EXERCISES

1.1 (a) Calculate the mean, median, and mode of each of the following four sets of readings:

(i) 4, 2, 7, 4, 5, 3, 5, 4, 3, 4.
(ii) 2, 1, 1, 3, 2, 2, 1, 1, 7, 1.
(iii) 0, 4, 5, 5, 0, 2, 5, 5, 0, 5.
(iv) 0, −2, 3, 0, 1, −1, 1, 0, −1, 0.

(b) Briefly summarize each of the data sets in terms of the mean. For example: 'The mean is . . . and most of the readings (say 80 percent) are within such and such a distance from the mean.'

1.2 Calculate the mean, median, and mode of the two data sets below:

(i) 44, 55, 58, 62, 65, 67, 69, 70, 73, 74,
74, 74, 77, 79, 81, 83, 89, 93, 96, 99.

(ii) 54, 67, 46, 34, 71, 83, 59, 52, 41, 39,
53, 74, 61, 53, 50, 28, 46, 69, 46, 54.

Which measure was the quickest to calculate in each case? State why.

1.3 Group the readings in each set of Exercise 1.2 as 20–29, 30–39, etc. Recalculate the means, medians, and modes, taking 25, 35, 45, etc. as the typical value in each interval. What are the benefits and the disadvantages of such grouping?

1.4 What kinds of data can be well summarized by measures like the mean, median, and mode? Which of the three measures is most commonly used? Why?

1.5 Summarize the following readings:

5, 4, 3, 5, 54, 4, 2, 3, 1.

What is the likely nature of the outlier?

1.6 (a) If twenty readings have a mean of 30 and another five readings have a mean of 80, what is the weighted average of all twenty-five readings? What is the unweighted average of the two means?

(b) Comment on the statement 'An average can only be used if it represents the typical value of a set of readings.' Give examples.

1.7 (a) Given two sets of four readings with their means, medians, and modes, as below, can you calculate the mean, median, or mode of the combined set of eight readings?

	Mean	Median	Mode
1st set (n = 4)	3	3	3
2nd set (n = 4)	5	5	5

(b) Use the two data sets below to illustrate the difficulties in computing the median and mode for combined data sets.

	Mean	Median	Mode
1st Example			
A: 2, 3, 3, 4	3	3	3
B: 3, 5, 5, 7	5	5	5
A & B: 2, 3, 3, 3, 4, 5, 5, 7	4	3.5	3
2nd Example			
P: 1, 3, 3, 5	3	3	3
Q: 4, 5, 5, 6	5	5	5
P & Q: 1, 3, 3, 4, 5, 5, 5, 6	4	4.5	5

CHAPTER 2

Scatter

Scatter is a statistical term which refers to the spread of readings in a set of data. When quoting an average, like 'The average age of the class is 21,' the scatter of the individual ages must also be described, at least implicitly. Often a simple statement will do, like 'Most of the ages lie between 19 and 23.' But sometimes more precision is needed.

The scatter of a set of readings can be described in terms of how much the readings differ from the mean. This leads to summary measures like the mean deviation, the standard deviation, and the variance. An alternative method is to note the range of the readings, the difference between the highest and lowest readings.

These summary measures of scatter are good descriptive devices when the data have a hump-backed and roughly symmetrical distribution. With skew distributions they may not give an adequate picture of the data. Then tailor-made summaries of the scatter may have to be used (or theoretical models as discussed in Chapter 5).

2.1 Deviations from the Mean

If we want to describe the scatter of a set of readings, we can summarize their deviations from the mean. These are simply the differences between each individual reading in a set and the mean of the set.

To illustrate, the following numbers of employees arrived late at an office on five different weekdays:

$$11, 2, 5, 1, 6.$$

The mean is 5. To find the deviations from the mean one subtracts the mean from each reading. The five deviations from the mean of 5 are therefore

$$6, -3, 0, -4, 1.$$

The minus signs show where the original readings are less than the mean. If we add these deviations, they will total zero: the readings above the mean balance those below the mean. This provides a useful check on our arithmetic, but tells us nothing about the size of the deviations.

When measuring their size we are concerned with how big the deviations are, and not whether they are positive or negative. We have to regard −3 as being as big a deviation as +3. Hence we take what are called 'absolute' values: the values ignoring any minus signs. (Absolute values are usually symbolized by vertical lines, like $|x|$ or $|x - \bar{x}|$.) The absolute deviations are therefore

$$6, 3, 0, 4, 1.$$

2.2   The Mean Deviation

The mean deviation is the straight average of the absolute deviations from the mean. For n readings

$$\text{Mean deviation} = \frac{\text{Sum of absolute deviations from the mean}}{n}.$$

Sometimes this is called the 'Mean Absolute Deviation' or 'MAD' for short, or may be abbreviated as md. Algebraically we can write

$$\text{md} = \frac{\text{Sum} |x - \bar{x}|}{n}.$$

In our numerical example, the mean deviation is

$$\frac{6 + 3 + 0 + 4 + 1}{5} = 2.8.$$

The interpretation of the mean deviation is simple. It states that the observed readings (11, 2, 5, 1, 6) lie on average within 2.8 units of their mean of 5. Thus three of the five readings lie within about 5 ± 2.8, i.e. between about 2 and 8. The other two readings lie somewhat beyond these limits (at 1 and at 11).

The mean deviation is therefore a common-sense measure of scatter. But it has two drawbacks:
  (i) Calculating the individual deviations from the mean can be time-consuming with less simple data. (Subtracting a two-digit mean like 8.9 from a large number of readings like 4.6, 1.9, 13.7, 10.4, etc. is bothersome.)
  (ii) The mean deviation cannot be used in subsequent mathematical calculations (as we will see later with one or two examples).

None the less, the mean deviation communicates well and is simple to use when summarizing scatter in small sets of data.

2.3   The Variance and the Standard Deviation

Two other statistical measures of scatter are the variance and the standard deviation. They are closely related to each other because the standard deviation is the square root of the variance. As concepts, these two measures are less obvious than the mean deviation. But they are often easier to calculate numerically and to manipulate mathematically afterwards.

These measures are based on the notion of first squaring the deviations from the mean and then averaging them. This is another way of measuring the size of the deviations irrespective of sign, since squares of negative numbers are always positive.

In the data on late-comers the deviations from the mean of 5 and their squares are

$$\text{Deviations} \qquad\quad 6, -3, \;\; 0, \;\; -4, \;\; 1.$$
$$\text{Squared deviations} \quad 36, \;\; 9, \;\; 0, \;\; 16, \;\; 1.$$

Each squared deviation is a measure of how far the original reading is from the mean. But the squaring has a bigger effect on large numbers than on small ones, so more numerical emphasis is given to the larger deviations.

In algebraic notation, if x stands for any one of the observed readings, and $\bar{x}$ is the mean, the deviation of x from the mean is $(x - \bar{x})$. The square of the deviation of x from the mean is $(x - \bar{x})^2$.

The Variance

The average of the squared deviations from the mean is called the variance. Sometimes it is abbreviated to var(x), meaning the variance of x:

$$\text{var}(x) = \frac{\text{Sum of squared deviations}}{\text{No. of readings}} = \frac{\text{Sum}\,(x - \bar{x})^2}{n}.$$

For our numerical illustration, the sum of the five squared deviations from the mean is 62. The variance is therefore

$$\text{var}(x) = \frac{62}{5} = 12.4.$$

The Standard Deviation

The variance measures the scatter of the readings from their mean, but is expressed in 'squared units'. To return to the original units of measurement, one takes the square root of the variance. This gives a measure known as the *standard deviation*, or s.d. for short:

Standard deviation $= \sqrt{\text{variance}}$

$$= \sqrt{\left(\frac{\text{Sum of squared deviations}}{\text{No. of readings}} \right)}$$

$$= \sqrt{\left(\frac{\text{Sum } (x - \bar{x})^2}{n} \right)}.$$

In our example, the variance was $62/5 = 12.4$. Hence the standard deviation is $\sqrt{12.4}$ or about 3.5. This is in the original units of measurement, i.e. numbers of late-comers. (Note that the s.d. at 3.5 is larger than the mean deviation at 2.8. This is generally the case because the s.d. gives more emphasis to large deviations.)

The procedure may seem complex but it is not. There are four steps:

(i) Take the deviations from the mean.

(ii) Square them.

(iii) Take the average of the squared deviations (the variance).

(iv) Take the square root of that average (the standard deviation).

Sometimes this operation is referred to as obtaining the 'root-mean-square' of the deviations.

The standard deviation says that as a rough average the readings are about 1 s.d. from the mean. As with the mean deviation, one can consider the interval of ± 1 s.d. on either side of the mean. In our example, three of the five readings 11, 2, 5, 1, 6 lie within 1 standard deviation on either side of the mean of 5, i.e. between 1.5 and 8.5. The other two again lie somewhat beyond these limits, at 1 and 11.

Much the same occurs with any data set that has an approximately symmetrical hump-backed shape: over half the readings will lie within ± 1 s.d. from the mean and the rest somewhat further out. (We shall pin this down more precisely in Chapter 5.) Thus one can summarize symmetrical hump-backed data by quoting their mean and standard or mean deviation.

Skew and U-shaped Distributions

Even with skew or U-shaped distributions, over half the readings usually lie somewhere within the limits of one standard deviation on either side of the mean. Very few will lie more than *two* standard deviations away from the mean.

But with non-symmetrical data the deviations from the mean will also be non-symmetrical. With a reverse-J-shaped distribution, there will be many small negative deviations and a few large positive ones. Table 2.1 gives an example.

**TABLE 2.1   A   Skew   Distribution
of Ten Readings**

1, 1, 1, 1, 2, 2, 3, 4, 6, 33
Mean: 5.4

The ten deviations from the mean of 5.4 are

$$-4.4, \ -4.4, \ -4.4, \ -4.4, \ -3.4, \ -3.4, \ -2.4, \ -1.4, \ .6, \ 27.6.$$

The variance is therefore 87.0 and the standard deviation is about 9.3. Here all but one of the original readings lie well within one standard deviation on either side of the mean (i.e. between −3.9 and 14.7). But eight of the ten deviations are negative. A good description of the data must say this.

It would be misleading to summarize skew data by merely quoting the mean and the standard deviation, as if the distribution were symmetrical. A more detailed description is needed. But there is no standard procedure. We need to use either a tailor-made summary (like 'The mean is at 5 but most of the readings are at 1 and 2, with one at about 30.') or a theoretical formula or 'model' of the observed frequency distribution, as discussed in Chapter 5.

2.4 A Short-Cut Formula

There is a short-cut method for calculating the variance, and hence the standard deviation, which avoids working out the deviations from the mean:

$$\text{Short-cut formula for the variance} = \frac{\text{Sum } (x^2) - n(\bar{x})^2}{n}.$$

In the regular formula one first needs to find and write down the difference $(x - \bar{x})$ between each reading and the mean, which is laborious. Then one squares the differences and sums the squares, and averages the results.

With the short-cut formula we square the original readings directly. We then sum these squares, subtract a 'correction for the mean' (n times the square of the mean), and average the result by dividing by n, the number of readings. This formula is particularly handy if the mean has already been calculated.

For our five readings of 11, 2, 5, 1, 6 we have

$$\text{Sum } (x^2) = 11^2 + 2^2 + 5^2 + 1^2 + 6^2$$
$$= 121 + 4 + 25 + 1 + 36 = 187$$

and

$$n\bar{x}^2 = 5 \times 5^2 = 125,$$

so that

$$\text{variance} = \frac{187 - 125}{5} = \frac{62}{5} = 12.4,$$

which is exactly what we got before. In the same way the standard deviation is $\sqrt{12.4} = 3.5$ just as before.

A simple verbal version of the short-cut formula to remember is

$$\text{Variance of } x = \text{Mean } (x^2) - (\text{Mean } x)^2,$$

i.e. the mean of the squares minus the square of the mean.

The standard deviation is far less laborious to compute with this short-cut formula. But the mean must be calculated to enough digits to avoid rounding-off errors if the difference between Sum (x^2) and $n\bar{x}^2$ is small compared to the general size of the x^2s. All of the calculations can be done in one operation on most calculators, without writing down any intermediate steps. (Some electronic pocket-calculators use the formula to compute the standard deviation so one merely has to key in the observed readings.)

This formula is one of many cases where the standard deviation is mathematically simpler to deal with than the mean deviation, for which no such short-cut exists.

2.5   Other Properties of the Standard Deviation

There are some other features of the standard deviation which need to be noted, starting with a complication in the basic formula.

The Divisor $(n-1)$

The formula for the standard deviation is generally used in a slightly different form than described so far. Instead of dividing the sum of the squared deviations by n (implying a straight average), the sum is divided by $(n-1)$:

$$\text{s.d.} = \sqrt{\left\{ \frac{\text{Sum } (x-\bar{x})^2}{(n-1)} \right\}} \quad \text{instead of} \quad = \sqrt{\left\{ \frac{\text{Sum } (x-\bar{x})^2}{n} \right\}}.$$

The divisor $(n-1)$ makes almost no numerical difference to the calculated standard deviation if the number of readings is greater than about 20. Even when $n = 10$, as for the data in Table 2.1, using $(n-1)$ as the divisor only increases the value of the standard deviation by 5 percent, from 9.3 to 9.8. The difference is trivial in terms of what the standard deviation tells us about the spread of the readings, which is from 1 to 33.

There are theoretical reasons for using $(n-1)$ instead of n. These apply to more advanced analyses involving sample data and will be explained later. With the $(n-1)$ divisor the short-cut formula of Section 2.4 becomes

$$\text{Variance} = \frac{\text{Sum } (x^2) - n\bar{x}^2}{(n-1)}.$$

Algebraic Notation

The standard deviation is often denoted by the symbol s or $s(x)$, meaning the standard deviation of the variable x. The variance is usually referred to as s^2. Thus one may see expressions like

$$s^2 = \frac{\text{Sum } (x - \bar{x})^2}{(n-1)},$$

or

$$s = \sqrt{\left(\frac{\Sigma(x - \bar{x})^2}{(n-1)} \right)},$$

using the Σ symbol for 'Sum'.

The Relationship with the Mean Deviation

The standard deviation is always larger than the mean deviation, but the difference is usually not big. The two measurements tell roughly the same thing: the average size of the deviations from the mean. For hump-backed and roughly symmetrical distributions the standard deviation is about 1.25 times the mean deviation. For highly skew distributions there is no general rule-of-thumb.

The Coefficient of Variation

The coefficient of variation (or CV for short) is a different way of expressing the standard deviation. It shows the standard deviation as a percentage of the mean:

$$\text{Coefficient of variation} = \frac{\text{Standard deviation}}{\text{Mean}} \times 100,$$

or

$$= \frac{s}{\bar{x}} \times 100.$$

Multiplying by 100 turns the ratio $s/\bar{x}$ into a percentage.

In our numerical example, the mean is 5 and the standard deviation is 3.5, so the CV is .7 or 70 percent:

$$\text{Coefficient of variation} = \frac{3.5}{5} \times 100 = .7 \times 100 = 70\%.$$

This measure says that the readings lie on average within about 70 per cent of 5. If the mean were much higher—say 50—but the standard deviation still 3.5, the CV would be much smaller, at 7 percent.

This way of expressing scatter can be useful when comparing data with different orders of magnitude or different units of measurement. Suppose the average height of a set of plants is 2 inches with a standard deviation of .5, and their average weight is 8 ounces with a standard deviation of 4. Then the coefficient of variation tells us the plants vary relatively more in weight than in height, i.e.

$$\text{CV of heights} = \frac{.5}{2} \times 100 = 25\%,$$

$$\text{CV of weights} = \frac{4}{8} \times 100 = 50\%.$$

Whether it is better to use the coefficient of variation or the straight standard deviation depends on the nature of the data. The CV is preferable if different data sets show roughly the same value. This makes it simple to use. To illustrate, Table 2.2 gives the heights of a group of oak trees at five ages, from 6 months to 6 years.

TABLE 2.2   Approximately Constant Coefficient of Variation of the Heights of Trees of Different Ages

	Age of Trees				
	6 months	1 year	2 years	4 years	6 years
Mean height (in inches)	2	5	10	40	120
Standard deviation	.5	1.2	3	9	31
Coefficient of variation	25%	24%	30%	22%	26%

The mean heights increase 60-fold, from 2 inches at 6 months to 120 inches at 6 years. The standard deviations of the heights of different trees at the same age increase in about the same ratio, from .5 to 31. The coefficient of variation is therefore virtually constant at about 25 percent. This allows a simpler description of the scatter of the different data sets than quoting their standard deviations. If the latter are wanted for some purpose, they can always be recalculated from the mean and the CV. Thus the standard deviation at 6 months should be about 25 percent of 2 inches, or .5, and at 6 years about 25 percent of 120 inches, or 30.

2.6   The Range

The range is the difference between the highest and lowest values in a data set. It differs from the measures discussed so far in that it is not based on deviations from the mean. It is a common-sense measure, but less useful for routine analyses than one might at first expect.

In our earlier numerical example the five readings were

$$11, \ 2, \ 5, \ 1, \ 6,$$

so the range is 10, i.e. from 1 to 11. Since the data are roughly symmetrical about the mean of 5, the mean with the range gives a fair description of the detailed data. The readings should vary between about $5 - (10/2) = 0$ and $5 + (10/2) = 10$, which is broadly correct.

But for skew data, we need to know the two end-points separately. Thus for the ten readings

$$4, \ 4, \ 4, \ 4, \ 4, \ 4, \ 4, \ 5, \ 6, \ 11,$$

the mean is 5 and the range 7. But the readings vary from only 1 unit below the mean to 6 units above the mean. As usual, skew data are harder to summarize.

We often want to know the range, or alternatively the two end-points, for a set of data. None the less, the range is not widely used as a routine measure of scatter. It has several drawbacks:

(i) Its value tends to increase with the number of readings, n. This makes it difficult to use when comparing the scatter in data sets with different n.

(ii) It is laborious to calculate unless the readings are already arranged in order of size, or n is small.

(iii) It is difficult to manipulate mathematically.

(iv) It reflects directly only the two extreme readings.

(v) It is greatly affected by an outlier.

The last property, however, makes the range into a useful device for checking whether one's data contain exceptional readings, high or low. These are shown up dramatically by an unusually large range. Thus the ten readings in Table 2.1 earlier had a mean of 5.4, but a range of 32 because of the isolated 'outlier' of 33.

The Inter-Quartile Range

To avoid the range's complete dependence on the two extreme values in a data set, the inter-quartile range is sometimes quoted. This is the interval which contains the central 50 percent of the readings, or as near to that as one can get. It is calculated by taking the difference between the value which marks the *lower quartile* (below which the lowest 25 percent of the readings lie) and the value which marks the *upper quartile* (above which the highest 25 percent of the readings lie).

The inter-quartile range is, however, even more laborious to calculate than the range. All of the readings have to be ordered by size, and not merely the highest and lowest picked out. Nor is there a short-cut formula, as for the standard deviation.

In any case, for hump-backed and roughly symmetrical data the inter-quartile

range says much the same thing as the standard or mean deviation: that the
interval of ±1 deviation from the mean contains just over 50 percent of the readings. For skew data, more detail has to be given anyway.

Although not widely used, the inter-quartile range reflects what we commonly
do when informally describing observed scatter: one leaves out the more extreme
readings and says something like, 'Most of the readings lie between 10 and 23.'

2.7   Discussion

The purpose of having standard measures of scatter is to provide a simple and
objective way of summarizing the variation in a set of readings. This is
particularly useful if the number of readings is large or if there are many different
sets of data. The procedure is straightforward if the distribution is hump-backed
and roughly symmetrical. There are then several different measures which
effectively tell the same story but which differ in convenience.

The mean deviation is useful in describing fairly small sets of data and it is easy
to understand. The variance and its square-root, the standard deviation, are more
useful in detailed analyses. They are also easier to compute using the short-cut
formula. The range indicates the extreme values in the data but is not widely used
for detailed analysis.

For roughly symmetrical hump-backed distributions, the standard deviation is
generally about 25 percent larger than the mean deviation. Thus one measure can
readily be converted to the other. Both say that on average the readings are
roughly one such measure away from the mean.

For skew data there is usually no simple routine way of summarizing the
scatter by a single measure. More specific descriptions may have to be used.

The amount and kind of variation in a set of readings affects how we interpret
and use the data. Earlier we had the example of the number of late-comers to an
office on five weekdays:

$$11, \ 2, \ 4, \ 1, \ 6. \qquad \text{Mean} = 5.0.$$

The figures vary a good deal from day to day, possibly making it difficult to
manage the early morning work-load. If the figures had been

$$4, \ 7, \ 3, \ 6, \ 5. \qquad \text{Mean} = 5.0,$$

the incidence of late-comers would have been easier to allow for.

CHAPTER 2   GLOSSARY

Absolute deviations	Values ignoring any negative signs.
Absolute values	Values ignoring the minus signs.
Coefficient of variation (CV)	Standard deviation/mean, expressed as a proportion or percentage.

Inter-quartile range	The length of the interval excluding the 25 percent largest and the 25 percent smallest readings in a data set.
md	An abbreviation of mean deviation.
Mean absolute deviation (MAD)	Another term for the mean deviation.
Mean deviation	The straight average of the absolute deviations from the mean.
$(n-1)$	The divisor commonly used in calculating the variance and standard deviation.
Range	The difference between the highest and lowest readings in a data set.
Root-mean-square deviation	Another term for the standard deviation.
Scatter	A general term for the spread of readings in a set of data.
Standard deviation	The square-root of the variance.
s	A symbol for the standard deviation.
sd	An abbreviation of standard deviation.
s^2	A symbol for the variance.
$s(x)$	A symbol for the standard deviation of the variable x.
Variance or var(x)	The average of the squared deviations from the mean, calculated as Sum $(x - \bar{x})^2/(n-1)$.

CHAPTER 2   EXERCISES

2.1 Calculate the mean deviation, the variance, and the standard deviation for the two sets of readings below.

Set A   3, 1, 6, 3, 3, 2.   Mean: 3.0
Set B   6.1, 1.7, 3.8, 5.3, 1.9, 4.1.   Mean: 3.82 approx.

Use the variance formula Sum $(\text{Reading} - \text{mean})^2/(n-1)$. Which data set was easier to handle? Why?

2.2 Calculate the variances of Sets A and B using the short-cut formula of Section 2.5 and comment.

2.3 (a) For each data set in Exercise 2.1, what percentage of the readings lie within

  – one mean deviation on either side of the mean,
  – two mean deviations on either side of the mean?

  (b) What are the corresponding percentages for the standard deviation?

  (c) Do the two measures of scatter tell the same story?

2.4 (a) In computing the standard deviations in Exercise 2.1, calculate the effect of using $(n-1)$ rather than n as the divisor.

  (b) How big is the effect when $n = 5$, $n = 10$, $n = 20$, and $n = 100$?

2.5 (a) Calculate the mean deviation and the standard deviation of the skew distribution

4, 4, 4, 4, 4, 4, 4, 5, 6, 11.

What percentage of the readings lie within

− one mean deviation on either side of the mean,
− two mean deviations on either side of the mean?

(b) How would you summarize the readings below, using the mean and either the standard or mean deviation?

5, 4, 3, 5, 54, 4, 2, 3, 1.

2.6 Calculate the mean deviation and the standard deviation of the following set of readings:

58, 55, 44, 74, 65, 89, 69, 96, 73, 74,
79, 62, 77, 74, 81, 83, 67, 93, 70, 44.   *Mean*: 74.1

Which measure was quicker to calculate? How could you simplify the calculation of the mean deviation?

2.7 (a) Calculate the range and the inter-quartile range for the data set in Exercise 2.6. Comment on their ease of calculation.
(b) Compare the range with the interval ± 3 mean deviations from the mean, and the inter-quartile range with the interval ± 1 mean deviation from the mean. Comment on the measures in the light of your results.

CHAPTER 3

Structured Tables

Most data we see are in the form of tables. Instead of the readings coming in any order, they are arranged in rows and columns. Often each entry in a table is an aggregate, like a total, an average, or a percentage. In this chapter we consider the analysis of such tables. Averages and the measurement of scatter as discussed in the last two chapters are again useful.

3.1 Row and Column Averages

Row and column averages in a table can provide a focus to guide the eye when looking at the individual figures.

Table 3.1 gives an example. It sets out the average heights of some 5000 Ghanaian children measured in the 1960s. The table differentiates them by age, sex, race, and living conditions.

TABLE 3.1 Average Heights of 6- to 12-Year-old Boys and Girls in Ghana

(in inches)	Age in Years							Av.
	6	7	8	9	10	11	12	
BOYS								
Rural	45	47	49	51	52	54	56	51
Urban	45	47	48	50	52	54	56	50
Privileged Urban	48	49	51	54	55	57	58	53
Expatriate (White)	46	49	52	54	55	58	59	53
GIRLS								
Rural	45	47	50	51	53	55	56	51
Urban	46	48	50	51	53	56	58	52
Privileged Urban	47	48	51	54	58	53	62	54
Expatriate (White)	46	49	52	53	55	59	61	54
Average	46	48	50	52	54	56	58	52

When looking at such a table it helps to start with the marginal averages:

The averages at the bottom of the table show a steady progression in height of about 2 inches per year, from 46 to 58. With this in mind, we see that the figures in each row show much the same growth rate.

The averages on the right of the table are all fairly similar and close to the overall mean of 52. With this in mind we see that the figures in each column of the table are also similar to each other. To a first approximation, all the 6-year-olds have average heights of about 46 inches, all the 7-year-olds average heights of about 48 inches, and so on.

More detailed analysis also turns on averages. For example, inspection of the eight averages in the right-hand column of the table shows that Rural and Urban children of either sex tend to be 2 or 3 inches shorter than Privileged Urban or Expatriate ones.

The preceding illustration has shown how averages can act as a visual focus when looking at the more detailed figures in a table. Table 3.2 without row and column averages is not as clear. It would not be as easy to summarize these results or to communicate them to others.

TABLE 3.2   The Data of Table 3.1 without Marginal Averages

		Age in Years						
(in inches)		6	7	8	9	10	11	12
BOYS								
Rural		45	47	49	51	52	54	56
Urban		45	47	48	50	52	54	56
Privileged Urban		48	49	51	54	55	57	58
Expatriate (White)		46	49	52	54	55	58	59
GIRLS								
Rural		45	47	50	51	53	55	56
Urban		46	48	50	51	53	56	58
Privileged Urban		47	48	51	54	56	58	62
Expatriate (White)		46	49	52	53	55	59	61

3.2   The Role of Averages

Row or column averages can also provide *summaries* of the data in a table. Whether they do so depends on the scatter of readings about the average. There are three main cases:

(i) When the readings are very similar.   An average provides a good summary if the figures in the row or column are all relatively close to each other (rather as for the hump-backed symmetrical distributions in Chapter 1). For example, in Table 3.1 we could say that the average heights of the various groups of 6-year-olds are all about 46 inches.

(ii) When the variation in different rows or columns takes the same form.   Here
the row or column averages can be atypical figures but still serve to differentiate
between the rows or columns, just as means can be used to compare skew
distributions which have the same shape. To illustrate, the row averages in
Table 3.1 are not typical of the individual figures in each row. But the variation
about the row averages takes a similar form: height increases by about 2 inches
per year. The row averages, therefore, help us to summarize the similarities or
differences between the rows: e.g. urban boys of all ages were of similar height to
rural ones, but expatriate and privileged boys were 2 to 3 inches taller.

*(iii) When the pattern of readings varies from row to row or from column to
column.*   Here averages provide no summary comparisons between different
rows or columns. It would be like comparing a reverse-J-shaped distribution with
a hump-backed one. But the averages can still provide a visual focus, to help us
see just how different the row or column patterns are.

An example of the third case is given by the energy data in Table 3.3.

TABLE 3.3   UK Coal and Oil Consumption

In giga therms	Coal	Oil	Av.
1960	24	12	18
'65	17	20	18
'70	12	27	19
'75	11	25	18
Average	16	21	18

The average consumption levels of 16 and 21 at the bottom of the table are not
typical of the individual figures in each column, nor of the difference between coal
and oil in most years. Similarly the *row* averages—all about 18—are not useful
summaries of the annual consumption levels of coal or oil nor of the differences
for either fuel from one year to another. Nonetheless, the averages act as a focus
helping us to see the more complex patterns inside the table. Relative to the constant row average of 18, coal is higher than oil in 1960 and lower in 1975.

Outliers

As usual, exceptional readings or 'outliers' in a table can complicate its description. In general it is best to omit outliers from the main averages and report them
separately.

3.3   Deviations from the Mean

The sub-patterns and exceptions in a table can be highlighted by looking at the deviations of the figures from the row or column averages, or at the deviations from both.

Table 3.4 shows the deviations from the column averages for the readings in Table 3.1. For example, rural boys aged 6 had an average height of 45 inches, so their deviation from the age-group mean of 46 is −1.

TABLE 3.4   Deviations from the Column Means of Table 3.1

(in inches)	6	7	8	9	10	11	12	Av.
			Age in Years					
BOYS								
Rural	-1	-1	-1	-1	-2	-2	-2	-1
Urban	-1	-1	-2	-2	-2	-2	-2	-2
Privileged Urban	2	1	1	2	1	1	0	1
Expatriate (White)	0	1	2	2	1	2	1	1
GIRLS								
Rural	-1	-1	0	-1	-1	-1	-2	-1
Urban	0	0	0	-1	-1	0	0	0
Privileged Urban	1	0	1	2	2	2	4	2
Expatriate (White)	0	1	2	1	1	3	3	2
Average (rounded)	0	0	0	0	0	0	0	0

Sub-patterns of relatively low or high figures in the original table show up more clearly as negative or positive deviations. In Table 3.4 rural girls and rural and urban boys of all ages generally score −1 or −2 and so are relatively short. Privileged and expatriate boys and girls score +1 or +2, and so are relatively tall.

Further inspection shows up more sub-patterns. The deviations for 10- to 12-year-olds look somewhat bigger than those for the 6- to 9-year-olds. We can also see that the deviations for 6- and 7-year-old *girls* are smaller than those for 6- and 7-year-old boys or for older children. Reference back to Table 3.1 shows the younger privileged and expatriate girls are almost no taller than rural or urban ones, unlike the rest of the data.

These various analyses are akin to the so-called 'Analysis of Variance' procedures sometimes used in more advanced statistical work, especially with small samples. These methods involve analysis of the total observed variation into main effects, interactions, and residuals. Systematic or general differences between row or column averages are called the *main effects* (e.g. the growth with age, or the overall rural/urban versus privileged/expatriate differences). More local patterns are called *interactions*. (An example is that girls are taller than boys for the older ages but the same difference does not hold for the younger children. So age and sex 'interact'.) Any variation that is left over is termed residual.

Interactions can be explored by calculating the deviations from both the column and row averages, as shown in Table 3.5. This gives a 0 for rural boys aged 6, showing that their being 1 inch below the average height of their age-group was typical for rural boys of *all* age-groups. Any value other than zero implies an interaction. For instance, the 1 for urban boys aged 6 implies that their height difference from other children aged 6 does not correspond with the differences for urban boys at other ages. We now see more clearly that the privileged and expatriate 6- and 7-year-old girls tend to be 1 or 2 inches shorter than other children of their age and type. The interaction effects here are numerically small. Larger ones would stand out more strongly in such a table of deviations. (The row and column averages are zero, subject to rounding errors.)

TABLE 3.5 Deviations from the Row and Column Means

Av. height	Age in Years							Av.
(in inches)	6	7	8	9	10	11	12	
BOYS								
Rural	0	0	0	0	-1	-1	-1	0
Urban	1	1	0	0	0	0	0	0
Privileged Urban	1	0	0	1	0	0	-1	0
Expatriate (White)	-1	0	1	1	0	1	0	0
GIRLS								
Rural	0	0	-1	0	0	0	-1	0
Urban	0	0	0	-1	-1	0	0	0
Privileged Urban	-1	-2	-1	0	0	0	2	0
Expatriate (White)	-2	-1	0	-1	-1	1	1	0
Average	0	0	0	0	0	0	0	0

Subtracting both column and row means and also any systematic interactions from the original readings leaves what is usually called the *residual variation*. (The results are the same whether one subtracts the column means or the row means first.) If there are no apparent or large patterns in these residuals, one can summarize them as being fairly irregular and of a specified average size.

The deviations in Table 3.5 in the main look irregular. They average at about $\frac{1}{2}$ inch, using the mean deviation as the measure of scatter. In more complex cases statisticians often use standard deviations or variances as measures of scatter, hence the name 'Analysis of Variance'.

Reporting the Results

The main task of statistical analysis is to summarize extensive data succinctly. The basic tool with structured tables is again averages. Thus in our example
  (i) Average growth is 2 inches per year.
  (ii) Privileged and expatriate children are on average $2\frac{1}{2}$ inches taller than rural and urban ones, but not for girls aged 6–8.

(iii) Black privileged children are on average about the same height as white expatriate ones, while non-privileged urban children are on average much the same as rural ones.
(iv) Older girls are on average about 1 inch taller than boys of the same age.
 (v) The residual deviations in the data are apparently irregular, averaging about $\frac{1}{2}$ inch.

3.4   Weighted Means and Totals

Row and column averages in a structured table do not take into account the number of individual readings that make up the base for each entry in the table. As a result, such averages may not be typical of all the individual readings.

For example, the average heights of rural and urban children and of privileged and expatriate ones were

Rural and urban   51.0 inches,
Privileged and expatriate   53.5 inches.

The straight average of the two figures is about 52.2 inches. But that is not the average height of all the 5000 children in the study. There were about 4500 rural and urban children and only 500 privileged and expatriate ones. The average height of all the 5000 children is therefore given by weighting the two means in proportion to the number of readings:

$$\frac{4500 \times 51.0 + 500 \times 53.5}{5000} = 51.3$$

This weighted mean of 51.3 is much closer to the mean of the rural and urban children, which makes sense since they greatly predominate in number.

But a weighted mean would only be needed if we wanted a single summary figure for the data as a whole. When there are large systematic differences in the data, like the 2-inch increase in height each year here, such an overall average is usually not a meaningful summary. There is no 'typical child' here; few children were anywhere near the overall average height. There is then little point in having a single overall average as a description, whether it is 51.3 (weighted) or 52.2 (unweighted).

In practice, the row and column means in a structured table are used mostly to help one see the patterns in the table, e.g. to summarize the differences between the different types of children and the way height increases with age. For such analytic purposes, unweighted row and column averages are simpler to use.

Totals

The weighted mean of a set of averages is the total of the individual readings divided by *n*. Totals can sometimes be physically meaningful in their own right, e.g. the total weight of a group of children when they go into an elevator. (But their total height is meaningless—'laid end-to-end . . .') In Table 3.6 the yearly totals of energy consumption have a physical meaning, and even perhaps the column totals.

TABLE 3.6   UK Coal and Oil Consumption

In giga therms	Coal	Oil	Total
1960	24	12	36
'65	17	20	37
'70	12	27	39
'75	11	25	36
Total	64	84	148

But for statistical analysis averages are generally far more helpful than totals since the averages are expressed on the same scale of measurement as the readings in the table. They therefore provide an easier focus or summary.

3.5   Discussion

Structured tables differ from frequency distributions of data in that each reading is differentiated systematically from another, e.g. boys' average heights versus girls', or last year's sales versus this year's.

The statistical tools already discussed in Chapters 1 and 2 are again useful with tables, i.e. averages and measures of scatter like the mean deviation. Although row or column averages often do not represent a typical value of the readings, they usually help us to see and communicate the main patterns in the data.

CHAPTER 3   GLOSSARY

Analysis of variance   A formal procedure for breaking the total variation in a structured table into main effects, interactions and residual variation.

Interactions   Cases where in the body of the table the differences between some rows are not the same for all the columns (the main effects 'interact').

Irregular deviations   Residual deviations which appear to show no systematic pattern.

Main effects   Differences in the marginal averages of a table from row to row or from column to column.

Residuals   The variation remaining in a table after allowing for systematic main effects and interactions.

Structured tables   A table arranged in rows and columns according to some conditions of observation.

Systematic difference   Or systematic variation: a consistent difference in the data.

CHAPTER 3   EXERCISES

3.1   (a)   The following table gives the consumption of five commodities over four quarters of the year. Briefly summarize the main features of the data.

lbs. per head	Quarters				Av.
	I	II	III	IV	
Potatoes	62	66	53	77	64
Flour	41	44	33	49	42
Vegetables	31	29	25	32	29
Meat	23	27	18	29	24
Fats	14	12	10	16	13
Average	34	36	28	41	35

(b)   Comment on the role of the averages.

(c)   Write out the deviations from the row averages. Do they help in analyzing the data?

3.2   (a)   Write in the row and column totals and averages for the following table and summarize the data.

(b)   Comment on your treatment of outliers and on the totals.

Quarter	Area				Av.	Total
	No	Ea	We	So		
1969 QI	98	75	50	48		
QII	92	75	57	42		
QIII	101	100	80	50		
QIV	90	74	51	39		
1970 QI	96	74	53	46		
QII	94	77	49	50		
QIII	91	72	59	42		
QIV	98	76	53	37		
Average						
Total						

3.3 The following numbers of persons died due to road accidents in five countries. Comment on the use of averages.

	1960	1965	1970
USA	38,000	49,000	56,000
Germany	14,400	15,800	16,000
France	8,300	12,100	14,300
Italy	8,200	9,000	9,800
UK	7,100	8,100	7,700

3.4 The table below gives the crop yields in an agricultural experiment. Three fields were used: one of high, one of medium, and one of low intrinsic fertility. In each field the yields were measured for two plots of ground applied with potash fertilizer, two plots with nitrogen fertilizer, two with both, and two with no fertilizer. Summarize the results.

Fertility	None	Potash	Fertilisers Nitrogen	Both
High	31, 37	48, 49	46, 50	47, 52
Medium	21, 28	37, 39	44, 44	45, 50
Low	8, 12	22, 24	25, 31	35, 45

3.5 The table below shows the percentages of viewers of each of the specified TV program who also watched the other programs. (Thus 49.0 percent of viewers of the 6 pm News also watched World in Action.) Use averages to help you describe the main patterns and exceptions in the data.

UK Programmes	Viewers of 6 pm News	Wld in Action	Tuesd. Film	Coron. Street	Adven- turers	News at 10
Who also Viewed	%	%	%	%	%	%
6 pm News	100.0	24.6	29.1	23.2	26.8	25.2
World in Action	49.0	100.0	49.8	54.2	51.3	47.6
The Tuesday Film	21.2	18.3	100.0	20.9	19.1	40.0
Coronation Street	53.6	63.4	68.5	100.0	59.1	54.7
The Adventurers	47.7	44.7	46.3	44.8	100.0	42.9
News at Ten	25.2	23.1	54.2	24.4	23.8	100.0

PART TWO:
FREQUENCY DISTRIBUTIONS

Frequency distributions have already been mentioned in Part One. They arise when we arrange a set of readings to see how often each value occurs. Chapter 4 discusses ways of handling such data.

Some frequency distributions can be described by theoretical formulae. This can be very convenient. Chapter 5 describes three of the more common examples: the Normal, the Poisson, and the Binomial distributions.

Chapter 6 introduces probability concepts. These can explain why certain types of frequency distributions occur.

CHAPTER 4

Observed Distributions

A frequency distribution sets out the number of times the different values occur in a set of readings. For example, it might show how many 0s, 1s, 2s, etc. occur for a measured variable. Or how many people have brown, blue, grey, or green eyes in a classification.

Figure 4.1 shows five of the most common distribution shapes in the use of measured variables. The different values are plotted along the horizontal axes and their frequencies are plotted vertically. The five shapes are

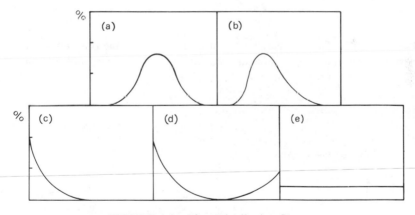

FIGURE 4.1   Five Distribution Shapes

(a) Approximately symmetrical with a peak or hump in the middle;
(b) Hump-backed but somewhat skew, i.e. with a relatively long tail of high (or low) numbers;
(c) Reverse-J-shaped and thus highly skew;
(d) U-shaped, i.e. with a mode at each end but one mode usually higher than the other;
(e) Uniform, with approximately equal frequencies for each observed value.

4.1   Counting the Frequencies

Observed readings often come in no particular order. The following example shows twenty students' test scores out of a maximum of 10:

$$6, \ 3, \ 7, \ 5, \ 6, \ 4, \ 4, \ 6, \ 7, \ 3, \ 5, \ 9, \ 6, \ 4, \ 2, \ 7, \ 5, \ 5, \ 8, \ 6.$$

The mean of 5.4 can easily be worked out by adding the readings (108) and dividing by $n = 20$. But to see the shape of the distribution the readings have to be *ordered by size:*

$$2, \ 3,3, \ 4,4,4, \ 5,5,5,5, \ 6,6,6,6,6, \ 7,7,7, \ 8, \ 9.$$

Now we can see that the distribution has a peak or mode roughly in the middle and that it is approximately symmetrical.

The number of times each value occurs in a set of data can be counted by setting out all the possible values in a row (or column), and recording the occurrence of each value:

	Observed values									
	1	2	3	4	5	6	7	8	9	10
Frequency	|	||	|||	||||		╫╫	|||	|	|	
	1	2	3	4		5	3	1	1	(20)✓

(If there is a great deal of data, such counting may be done on a computer.) To spot errors, one must check that the total number of readings and their sum (or mean) agrees with the original set of readings, e.g. that $n = 20$ and that the total is 108.

The layout is sometimes more formal, as in Table 4.1.

TABLE 4.1   **The Frequency Distribution of Twenty Test Scores**

	Test Scores		Total no. of students
	0 1 2 3 4 5 6 7 8 9 10		
Frequency:	- - 1 2 3 4 5 3 1 1 -		20✓

To calculate the mean of a frequency distribution in this form, we multiply each measured value by its frequency of occurrence, sum these products, and divide the total by n. In our example this gives:

$$0 \times 0 + 1 \times 0 + 2 \times 1 + 3 \times 2 + 4 \times 4 \times 3 + \ldots + 9 \times 1 + 10 \times 0 = 108,$$

which divided by $n = 20$ gives $\bar{x} = 5.4$. If each observed value is denoted by x and its frequency by f, the mean can be expressed algebraically as

$$\bar{x} = \frac{\text{Sum}\,(x \times f)}{n}.$$

The same form of calculation is used to find the standard deviation of a frequency distribution:

$$\text{Standard deviation} = \sqrt{\frac{\text{Sum}\,\{(x - \bar{x})^2 f\}}{(n - 1)}}.$$

4.2   Relative Frequencies

It is easier to compare different frequency distributions if we show the numbers as *relative frequencies*. These express the observed frequencies as proportions of the total number of readings in each set.

To illustrate, Table 4.2 gives two sets of readings: our 20 students' test scores and those of another class of 50 students. The shape of the two distributions is clearly similar. But detailed comparisons are not easy because the total numbers of readings differ.

TABLE 4.2   Two Sets of Readings to Compare

	Observed Value											Total no. of readings
	0	1	2	3	4	5	6	7	8	9	10	
1st Set	-	-	1	2	3	4	5	3	1	1	-	20
2nd Set	-	-	3	6	7	13	10	7	3	1	-	50

The solution is to express the observed frequencies as relative frequencies, dividing through by $n = 20$ and $n = 50$. (Percentages are often easier to handle than proportions and so we divide each observed frequency by n and multiply by 100.) The similarities and the differences between the two sets of data are now clearer, as shown in Table 4.3.

TABLE 4.3   The Frequency Distribution as Percentages

	n		Observed Value									
			0	1	2	3	4	5	6	7	8	9	10
1st Set	20	%	-	-	5	10	15	20	25	15	5	5	-
2nd Set	50	%	-	-	6	12	14	26	20	14	6	2	-

To calculate the mean of such a distribution, we multiply each value by its relative frequency, sum these products, and then divide by 100. (The denominator is

100 because the percentages refer to a base of 100 percent.) Thus the first set has a mean of 5.40 and the second a mean of 5.18.

4.3   Grouped Data

To reduce the size of tabulations and also smooth over gaps and irregularities in the data, adjacent readings in a frequency distribution are often grouped together. Table 4.4 illustrates this by grouping most of the readings of the previous table into two-unit intervals like 3–4, 5–6, etc, plus 'open-ended' intervals like −2 and 9+.

TABLE 4.4   Grouped Categories

			Observed Value			
		-2	3-4	5-6	7-8	9+
1st Set	%	5	25	45	20	5
2nd Set	%	5	26	47	20	2

Some grouping is inevitable when summarizing continuous measurements, like giving people's ages to their last birthday rather than to the nearest month or minute. But it can be used with any data. The choice of grouping intervals is often determined by habit. For instance, we often see ages shown in 5- or 10-year groupings (but rarely in 6-year ones), or in still broader and unequal categories like '0–17', '18–24', '25–44', '45–64', and '65 + '.

Data grouped into 4 to 6 categories are usually easier to read than long strings of numbers. But grouping handicaps the calculation of summary measures like the mean or variance, as there is no precise value for each grouping interval. Midpoint estimates are often used, like 3.5 for the 3–4 interval in Table 4.4. The shape of the distribution may however imply that there were more 4s than 3s, so that a value like 3.7 or 3.8 might be more accurate. Imprecision can arise particularly with 'open-ended' categories at the extremes of a distribution. When these occur it may be worthwhile to list the individual readings.

4.4   Percentiles

If we add the frequencies of readings from the lowest value on the left of a distribution, we obtain *cumulative* frequencies. These tell us what percentage of readings are equal to or less than a given value.

Table 4.5 shows the relative frequencies and corresponding cumulative frequencies for the 20 test scores. Thus of the 20 students, 5 percent got a score of 2 (i.e. 1 student), 15 percent got 3 or less, 30 percent got 4 or less, and 100 percent got 9 or less. Cumulative frequencies are like progressive grouping.

TABLE 4.5   The Relative and Cumulative Frequencies for the Twenty Students' Test Scores

		Test Scores											Total
		0	1	2	3	4	5	6	7	8	9	10	
Relative Frequency	%	0	0	5	10	15	20	25	15	5	5	0	100
Cumulative Frequency	%	0	0	5	15	30	50	75	90	95	100	100	100

Figure 4.2 plots the cumulative frequencies on a graph. For hump-backed symmetrical distributions the picture takes what is called an 'S-shape' with a fairly slow start, a fast build-up of frequencies in the middle range, and a slowing down to reach 100 per cent.

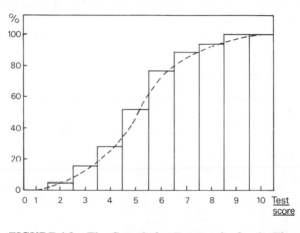

FIGURE 4.2   The Cumulative Frequencies for the First Set of Data

Cumulative frequencies show what percentage of readings are equal to or less than any given value, e.g. that 75 percent of the readings in Table 4.5 are 6 or less. Such a value is referred to as the 75th *percentile* of the distribution. The 50th percentile, the value below which 50 percent of the readings lie, is the median. The 10th percentile or 'lower decile' is the value below which 10 percent of the readings lie and the 90th percentile or 'upper decile' is the value above which 10 percent of them lie. In our example the 50th percentile is at 5, the lower decile is between 2 and 3 and the upper decile is at 7.

The *range* is the difference between the 0th and 100th percentiles, so it covers 100 percent of the readings. A *quartile* is a value separating off a quarter or 25 percent of the readings. Thus the *inter-quartile range* of Section 2.6 is the difference between the observed values at the 25th and 75th percentiles, the limits containing the central 50 percent of the readings.

Only selected values of cumulative frequencies need to be given when reporting data. For example, with our two data sets, every other value could be omitted without much loss of information but greatly improved clarity:

	Values			
	−3	−5	−7	−9
Cumulative Frequencies				
First Set (%)	15	50	90	100
Second Set (%)	18	58	92	100

By giving less detail we are communicating more.

Cumulative frequencies and the percentages of readings that lie between two particular values are the common language for describing frequency distributions. For example, we could summarize the data in Table 4.3 quite well by saying that roughly 70 percent of the readings in each set lie in the interval from 4 to 7.

4.5 Graphs and Histograms

Plotting a frequency distribution on a graph helps to show its shape. The observed values are usually plotted horizontally while the actual or relative frequencies are plotted vertically. Figure 4.3 shows a graph of the following data:

Value	1	2	3	4	5	6
Relative frequency (%)	5	15	30	40	8	2

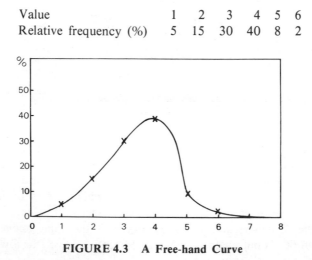

FIGURE 4.3   A Free-hand Curve

To see the shape of the distribution more clearly, a smooth curve has been drawn through the observed points. Such a curve can be used even if the measured variable is discontinuous, i.e. if it takes only discrete or distinct values, like whole numbers.

With grouped data, one often plots *histograms*. They are diagrams showing rectangular bars instead of curves. The *area* of each bar represents the frequency of values in that particular grouping interval. This device allows one to use unequal intervals. To determine how high each bar should be, the frequency of the group is divided by the width of the grouping interval. Figure 4.4 illustrates a histogram for the following grouped version of the data:

Value	1–2	3	4	5–6
Relative frequency (%)	20	30	40	10

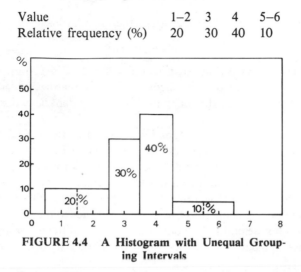

FIGURE 4.4 A Histogram with Unequal Grouping Intervals

The width of each interval is usually taken symmetrically about the plotted value. For example, the frequency of 3 is 30 percent, so the bar is given a height of 30 and a width of 1, plotted from 2.5 to 3.5. But the 20 percent frequency of the 1–2 grouping is plotted at a height of only 10 percent, with the bar going from .5 to 2.5. This gives 10 percent for the '1' interval and 10 percent for the '2' interval, thus sharing the 20 percent equally among the constituent units in the grouping interval. If all the intervals are equal, a simple graph showing the frequencies as heights looks the same as a histogram.

4.6 Standardized Variables

Sometimes it is helpful to turn the readings in a frequency distribution into a 'standardized variable'. To do this one first subtracts the mean $\bar{x}$ from each observed reading and then divides the resulting differences by the standard deviation of the readings. This gives the new variable a mean of zero and a standard deviation of 1. In algebraic notation the standardized form of variable x reads

$$\frac{x - \bar{x}}{s}.$$

As an example, consider the five readings

$$2, 4, 5, 5, 9.$$

They have a mean of 5.0 and a standard deviation of about 2.5. If we subtract the mean $\bar{x} = 5$ from each reading, we get

$$-3, -1, 0, 0, 4.$$

Dividing each figure by $s = 2.5$ then gives the 'standardized' readings

$$-1.2, -.4, 0, 0, 1.6.$$

These readings should have a mean of 0 and a standard deviation of 1 (within rounding errors), as can readily be checked. The standardized reading 1.6 for example tells us that the original reading (9) lies 1.6 standard deviations above the original mean.

Standardized variables are used in dealing with certain theoretical distributions (as we shall see in Part Three). They can also aid comparisons of different sets of measurements, like different pyschological test scores for the same group of people. The device expresses all the variables on a common scale, e.g. one with a mean of 0 and a standard deviation of 1. (In psychology IQ scores are often standardized with a mean of 100 and a standard deviation of 15.) However, standardization is not helpful when comparing data for different groups of people or items because it hides any differences in the observed means and standard deviations.

4.7   Discussion

The shape of a frequency distribution conveys basic information about the data, e.g. whether it is hump-backed or U-shaped. Summary measures like an average or a standard deviation do not mean much unless the distribution's shape is known. For example, an average will be misleading if the distribution is highly skew but assumed to be symmetrical.

Summarizing a set of readings therefore usually involves describing four main aspects of the data:

(i) The number of readings, n.
(ii) Their average size, e.g. the mean m.
(iii) Their scatter, e.g. the standard deviation s.
(iv) The distribution shape.

When dealing with many different data sets, information about the numbers of readings and the shape and scatter of each distribution is often unstated. This typically occurs with tables of averages or totals (as for example discussed in Chapter 3). The practice is acceptable if the information is at least implicit or the details are not crucial (e.g. that all the distributions are of the same shape and that none of the *n*s are tiny).

If different data sets have the same general shape and similar amounts of scatter, we can in fact compare them simply by using their means. This can be so even for skew distributions, where the mean is not typical of the bulk of the readings. Such use of the mean occurs with many of the mathematical models discussed in Chapter 5.

CHAPTER 4   GLOSSARY

Continuous measurements	Measurements which can take any numerical value in a certain range.
Cumulative distribution	A distribution showing the cumulative frequencies in a set of readings.
Cumulative frequencies	A step-by-step summation of the frequencies, starting from the lowest.
Discontinuous measurements	Variable measurements which can only take certain values, like whole numbers.
Discrete variable	Another name for discontinuous measurements.
Frequency distribution	A count of the number of times each value occurs in a set of readings.
f	Symbol denoting the frequency of a value in a set of readings.
Grouped data	Data where adjacent readings have been combined into groups.
Histogram	Bar-chart showing the frequency of readings, used especially with grouped data.
Inter-quartile range	The difference between the observed values at the 25th and 75th percentiles of a distribution.
Lower decile	The 10th percentile of a distribution.
Percentile	The observed value which relates to a particular cumulative frequency, e.g. the median or 50th percentile has 50 percent of the readings lying below it.
Quartile	The value which marks off one-quarter of the readings, as at the 25th or 75th percentiles.
Relative frequency	A frequency expressed as a proportion or percentage of the total number of readings.
Standardized variables	Measurements which have had their scale changed to give them a mean of 0 and a standard deviation of 1.
Upper decile	The 90th percentile of a distribution.

CHAPTER 4   EXERCISES

4.1   (a)   Arrange the following forty readings as a frequency distribution:

7, 4, 6, 2, 5, 5, 8, 3, 6, 6, 5, 4, 7, 7, 8, 3, 6, 9, 5, 7,
5, 3, 6, 8, 6, 3, 5, 7, 4, 3, 3, 5, 6, 5, 7, 4, 6, 5, 5, 4.

(Check the number of readings and the mean of the frequency distribution with the original data.)

(b) To count two-digit numbers one can use a two-dimensional grid, counting the first digit vertically and the second digit horizontally. Use the following grid to construct a frequency distribution of the numbers

$$43, \ 29, \ \ 0, \ 9, \ 46, \ 61, \ \ 0, \ 13, \ 17, \ 4,$$
$$15, \ \ 0, \ 51, \ 2, \ 54, \ 33, \ 20, \ 18, \ \ 6, \ 0.$$

For example, 43 gets a 'stroke' in the fourth row and the third column:

Tens	\ \ \ \ \ \ \ \ \ \ Units									
	0	1	2	3	4	5	6	7	8	9
0										
1										
2										
3										
4			*I*							
5										
6										

4.2 (a) Take the two distributions in Exercise 4.1 and group each into four to six suitable categories.

(b) Using mid-interval values work out the mean and standard deviations of the first grouped distribution. Compare them with the corresponding values of the ungrouped data.

4.3 (a) When is it useful to express observed frequencies as relative frequencies?

(b) Write the two sets of data in Exercise 4.1 as cumulative frequency distributions and also express them as percentages.

4.4 (a) Draw a suitable histogram of the distribution in Exercise 4.1(a).

(b) On the same graph superimpose a histogram of the *grouped* version of the data from Exercise 4.2. Explain why this gives similar-looking results, even though the grouped frequencies are higher.

4.5 For the data in Exercise 4.1(a), express '7' as a standardized variable. Explain its meaning in words.

CHAPTER 5

Theoretical Distributions

Some observed frequency distributions can be described by mathematical formulae. Such theoretical models then provide a convenient shorthand for summarizing the data.

One example is called the *Normal* distribution. This often gives quite a close description of hump-backed symmetrical data, like the distributions of people's heights or IQs. Skew distributions—as occur when counting how many car journeys people take or how many accidents they have—can sometimes be described by Poisson, Binomial, or related distributions. But one cannot always find a suitable theoretical model to describe observed data.

While the mathematical formula for a theoretical distribution can be fairly complicated, in practice only a simple numerical table may have to be used. Computers can also help to simplify the technicalities. The later sections of this chapter contain a good deal of detailed information. On a first reading it is enough to appreciate the variety of formulae which can be useful in modelling or describing observed data.

5.1 The Normal Distribution

The Normal distribution is a mathematical formula which describes data that follow a particular type of hump-backed and symmetrical frequency distribution. Its shape is shown in Figure 5.1, which indicates the relative frequency of readings

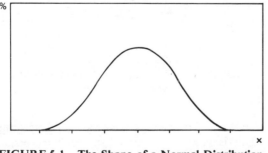

FIGURE 5.1   The Shape of a Normal Distribution

at any particular value of *x*. The mean, median and mode of a Normal distribution are equal. (This distribution is sometimes also called the *Gaussian* distribution, after the nineteenth century German mathematician Gauss.)

In a Normal distribution the frequency with which a value *x* occurs is proportional to the expression

$$e^{-(x-\mu)^2/2\sigma^2}$$

Here *e* is a mathematical constant equal to 2.718; μ is the mean (Greek m pronounced 'mu'); and σ (Greek s pronounced 'sigma') is the standard deviation of the distribution. (Greek letters are widely used for theoretical models.) The frequency of *x* depends on how far *x* is from the mean μ, relative to the standard deviation σ.

The formula looks complicated but no direct use of it has to be made in practical work. Instead one can generally refer to the percentage of readings which lie between two points equally distant from the mean. Usually just three such percentages will be enough, namely that in a Normal distribution

> 68% of the readings lie between $\pm 1\sigma$ from the mean,
> 95% of the readings lie between $\pm 2\sigma$ from the mean,
> 99.7% of the readings lie between $\pm 3\sigma$ from the mean.

Thus saying that some data is Normal with a certain mean and standard deviation is often a good enough description. It tells us what the data are like.

Figure 5.2 illustrates these ranges for a Normal distribution with a mean of 9 and a standard deviation of 3: 34 per cent of the readings lie between 6 and 9 and another 34 percent lie between 9 and 12. This makes a total of 68 percent between 6 and 12, i.e. between ± 1 standard deviation σ from the mean.

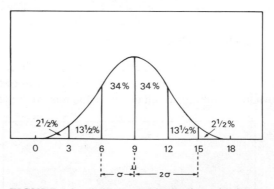

**FIGURE 5.2   A Normal Distribution with Mean
9 and Standard Deviation 3**

Similarly, 2.5 percent of the readings are less than 3 and another 2.5 percent greater than 15. This leaves 95 percent of the readings inside the $\pm 2\sigma$ limits, between 3 and 15. Only 0.3 percent of the readings, i.e. almost none, lie more than ±3σ away from the mean. Occasionally other percentages are also required. Table 5.1 gives a wider selection. (See also Table A1 in Appendix A.)

TABLE 5.1   The Descriptive Properties of the Normal Distribution
(For selected values of the standard deviation σ)

% of readings Within or Outside the stated limits	± distance from the mean								
	.5σ	.3σ	1σ	1.6σ	2σ	2.6σ	3σ	3.3σ	3.9σ
Within :	% 40	% 58	% 68	% 90	% 95	% 99	% 99.7	% 99.9	% 99.99
Outside:	60	42	32	10	5	1	.3	.1	.01

Theoretically the Normal distribution ranges from minus infinity to plus infinity. However, the proportion of readings lying way out is tiny. The table shows how only 1 reading in 10 000 lies more than 3.9σ away from the mean. Thus the distribution can be used to describe measurements which lie within a restricted range (e.g. ones which cannot take negative values).

The Normal distribution is *continuous*. This means the variable can take all numerical values, like weights, distances, and temperatures do for example. (Strictly speaking, the mathematical formula refers to the proportion of readings lying in a particular small interval.) The distribution can also be used as an approximation with discrete variables—ones that take only fractional or whole values, as when things are being counted.

A special property of Normal distributions is that they always take the same kind of shape, no matter what their mean and standard deviation. Figure 5.3 illustrates that 68 percent of the readings lie between ±1σ from the mean, whatever the value of σ or μ. This makes the distribution exceptionally simple to use.

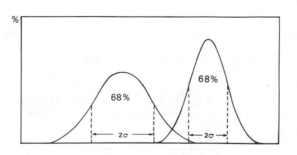

FIGURE 5.3   Two   Normal   Distributions   with Different Means and Scatter

The mean deviation (m.d.) of Chapter 2 can also be used to describe the scatter of the Normal distribution, e.g. that

58% of the readings lie between ± 1 m.d. from the mean,
90% of the readings lie between ± 2 m.d. from the mean,
98% of the readings lie between ± 3 m.d. from the mean.

For a Normal distribution, the mean deviation is about 20 percent lower than the standard deviation. Thus

1 mean deviation $\doteqdot$ 0.8 standard deviation, or
1.25 mean deviation $\doteqdot$ 1 standard deviation,

where $\doteqdot$ means 'approximately equal'.

To see whether a Normal distribution fits observed data we can check, (i) whether the readings are approximately symmetrical (half above the mean and half below), and (ii) whether about 68 percent of the readings lie between the $\pm 1\sigma$ limits and about 95 percent between the $\pm 2\sigma$ limits (or between the corresponding mean deviation limits). The fit will seldom be perfect. It is judged good if different sets of data do not show the same pattern of deviations (e.g. consistently more than 5 percent of the readings outside the $\pm 2\sigma$ limits). The data are then easy to describe, i.e. as approximately Normal (with irregular deviations).

Another method for checking the fit of the model is to plot the observed data as a cumulative distribution on 'Normal probability paper'. With a good fit the observed distribution shows up simply as a straight line. This is mentioned here because statistical computer packages often contain this procedure. But for most purposes it is unduly elaborate compared with checking out the $\pm 1\sigma$, 2σ, and possibly the 3σ percentages.

Numerical Example

The Normal distribution gives such a close description of so many observed phenomena that it is important to be familiar with its role in summarizing and comparing different data sets.

Suppose we have blood sugar readings for $n = 20$ patients, as shown in Table 5.2. The readings have a mean of 5.4 and a standard deviation of 1.8.

TABLE 5.2   The Relative Frequency Distribution of Twenty Readings

	n	Observed Value															Average
		0	1	2	3	4	5	6	7	8	9	10	11	12	13	14	15	
1st Set	20	%	-	-	5	10	15	20	25	15	5	5	-	-	-	-	-	-	5.4

Stand. dev. = 1.8

Half the readings are above the mean and half below and the distribution looks broadly symmetrical. Seventy-five percent rather than 68 percent lie between $\pm 1\sigma$ from the mean, and 95 percent or 100 percent lie within $\pm 2\sigma$ (depending on whether one counts the reading of 9.0). With a base of only twenty readings, where one reading amounts to 5 percent of the total number, the fit for a Normal distribution looks almost as close as it can be.

Saying that the twenty readings follow a Normal distribution with mean $= 5.4$ and s.d. $= 1.8$ is therefore true to a close degree of approximation.

The same is true for the second set of blood-sugar readings shown in Table 5.3. These are for $n = 80$ older patients and have a mean of 8.3 and a standard deviation of 1.8.

TABLE 5.3   Two Sets of Data

	n	Observed Value															Average
		0	1	2	3	4	5	6	7	8	9	10	11	12	13	14 15	
1st Set	20	%	-	-	5	10	15	20	25	15	5	5	-	-	-	-	- -	5.4
2nd Set	80	%	-	-	-	1	-	6	8	17	20	24	14	7	2	1	- -	8.3

The two distributions can therefore be summarized as having about the same shape and scatter, i.e. Normal with an s.d. of 1.8. They differ only in their means. This summary is set out formally in Table 5.4.

TABLE 5.4   Summary of the Two Sets of Data

	n	Distribution (Approx.)	Mean	Standard Deviation
1st Set	20	Normal	5.4	1.8
2nd Set	80	Normal	8.3	1.8

Different distributions that follow the same theoretical formula are differentiated in terms of certain characteristic values or *parameters*. The Normal distribution has two parameters. These are generally denoted by the mean μ and the standard deviation σ, but other characteristic values will do, like the mean deviation instead of the standard deviation.

Non-Linear Transformations

Many observed distributions are skew, and thus complex to describe. Sometimes the scale of measurement can be changed to make the data more symmetrical, and indeed Normal in shape. Hence the data become easier to summarize.

To illustrate the process we consider a square-root transformation. Suppose some measure of the size S of fifteen bacterial cultures after incubation is

$$S: \quad 0, \quad 1.1, \quad 3.7, \quad 9.8, \quad 17.1,$$
$$23.6, \quad 26.4, \quad 35.5, \quad 36.4, \quad 48.2,$$
$$65.7, \quad 82.3, \quad 103, \quad 124, \quad 174.$$

The mean is 50, with ten out of the fifteen readings less than this. The distribution is therefore rather skew, as Figure 5.4A shows.

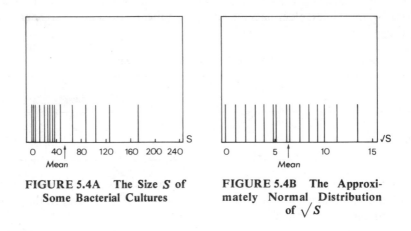

FIGURE 5.4A   The Size S of Some Bacterial Cultures

FIGURE 5.4B   The Approximately Normal Distribution of $\sqrt{S}$

If we take square-roots of the readings, we get the following fifteen numbers (to the nearest digit):

$$\sqrt{S} \quad 0, \ 1, \ 2, \ 3, \ 4, \ 5, \ 5, \ 6, \ 6, \ 7, \ 8, \ 9, \ 10, \ 11, \ 13.$$

The distribution has become broadly symmetrical, as shown in Figure 5.4B. The mean is 6 and the mean deviation is 2.9. Seven of the fifteen readings (47 percent) lie within ± 1 m.d. from the mean and 13 out of 15 (87%) within ± 2 m.d. Thus the distribution of $\sqrt{S}$ is fairly close to Normal.

The transformed data are now simpler to describe: the variable $\sqrt{S}$ is approximately Normal, with a mean of 6 and a mean deviation of 2.9. This still tells us what the original data were like:

About 58 percent of the $\sqrt{S}$ readings should lie between about 3 and 9 on the $\sqrt{S}$ scale. Therefore about 58 percent of the S readings should lie between $3^2 = 9$ and $9^2 = 81$ on the original S scale. Reference back to the S readings shows that in fact 9/15 or 60 percent do.

Similarly, about 90 percent of the readings should lie between 0 and 12 on the $\sqrt{S}$ scale. Therefore about 90 percent of the S readings should lie between 0 and 144 on the S scale, as is the case again.

The most common transformation is to take *logarithms* (see Table A5 in

Appendix A). This has a more drastic effect than square-roots, with larger values being brought still closer together:

$$
\begin{array}{llllll}
x & 1 & 10 & 100 & 1000 & 10\,000 \\
\sqrt{x} & 1 & 3 & 10 & 32 & 100 \\
\log x & 0 & 1 & 2 & 3 & 4
\end{array}
$$

A physical explanation for using logarithmic transformations is provided by the notion of exponential growth in biology and economics. For example the size of some organisms tends to multiply in successive units of time, from 1 to 2 to 4 to 8 to 16 to 32, etc. (rather than growing by a fixed additive amount, from 1 to 3 to 5 to 7, etc.). In such cases the log-sizes of different organisms at a given point in time tend to be distributed Normally. The original (skew) readings are said to follow a *lognormal* distribution.

Finding an appropriate mathematical function to transform skew distributions is generally a matter of informed trial and error. Some observed distributions are approximately symmetrical but differ consistently from the Normal in the proportion of readings in the extreme tails, e.g. 8 percent beyond $\pm 2\sigma$, instead of about 5 percent. This is referred to as *kurtosis*. There are no general procedures for dealing with such data and they have to be described by tailor-made summaries.

5.2 The Poisson Distribution

Data which come as counts rather than as continuous measurements are often very skew. Poisson or related theoretical distributions can sometimes be used to describe such data.

The Poisson distribution, named after an eighteenth century French mathematician, assumes four things about the items or events being counted: (a) that they have no effective upper limit; (b) that they occur independently of each other; (c) that they occur irregularly; and (d) that the average rate or probability of occurrence is constant.

Table 5.5 shows an example where the Poisson distribution gives a good fit to observed data. The readings concern the commencement of major labor strikes in the United Kingdom over 12 years. There were 563 strikes, an average of .9 strikes per week. The observed data are closely matched by a theoretical Poisson

TABLE 5.5 The Fit of the Poisson Distribution for the Occurrence of 'Major Strikes' in the UK per Week in 1948–1959

		Strikes per week						Total number of weeks
		0	1	2	3	4	5+	
Number of Weeks	Observed	252	229	109	28	8	0	626
	Poisson	255	229	103	31	7	1	626

distribution with a mean of .9. (The largest discrepancy is for the number of weeks with two strikes.)

One benefit of this close fit is descriptive: the data can be summarized simply by saying 'It's approximately Poisson with a mean of .9.' A second benefit is interpretative: since the underlying mechanism of the Poisson model assumes independent occurrence, the fit is consistent with the idea that one strike did not trigger off another.

The Mathematical Formula

A Poisson distribution has only one characteristic value or parameter: its mean μ. Given the value of μ, the rest follows. A special mathematical property of the Poisson is that its variance is always numerically equal to its mean:

$$\text{mean} = \mu = \text{variance}.$$

In a Poisson distribution the proportion p_r of the readings which takes the value r is given by

$$p_r = \frac{\mu^r e^{-\mu}}{r!}.$$

This is not as complicated as it looks. The symbol ! is called *factorial*. It is a shorthand way of saying that one has to multiply the specified number by every positive whole number below it. Thus $r!$ stands for the product $r \times (r-1) \times (r-2) \times \ldots \times 3 \times 2 \times 1$. For example 4! is $4 \times 3 \times 2 \times 1 = 24$. (The factorial 0! is defined to equal 1. So is any number raised to the power of zero, i.e. $\mu^0 = 1$. Hence $p_0 = e^{-\mu}$.)

TABLE 5.6   Selected Values of e^{μ}

μ	$e^{-\mu}$	μ	$e^{-\mu}$	μ	$e^{-\mu}$
.01	.990	.1	.905	1	.368
.02	.980	.2	.818	2	.135
.03	.970	.3	.741	3	.050
.04	.961	.4	.670	4	.018
.05	.951	.5	.607	5	.007
.06	.942	.6	.549	6	.002
.07	.932	.7	.497	7	.001
.08	.923	.8	.449	8	.000
.09	.914	.9	.407	9	.000

<u>Example:</u> To calculate $e^{-.61}$,

write $e^{-.61} = e^{-.6} \times e^{-.01}$

$= .549 \times .990$

$= .544$

The expression $e^{-\mu}$ involves the same mathematical constant $e = 2.718$ that occurs with the Normal distribution. Values of $e^{-\mu}$ for different values of μ can be found in Table 5.6.

To illustrate the calculations suppose we want to check whether the observed data on strikes in Table 5.5 follow a theoretical Poisson distribution. First we see whether the variance of the observed data is equal to the mean. The variance is .88, following the usual variance calculations in Chapter 2. This is close to the mean of .90, so it is worth continuing.

Next we use the mathematical formula for p_r. For the proportion of weeks with three strikes, for example, we insert $\mu = .9$ and $r = 3$ into the formula for p_r and get

$$p_3 = \frac{.9^3 e^{-.9}}{3!}.$$

We compute 3! as $3 \times 2 \times 1 = 6$. Table 5.6 shows that $e^{-.9} = 407$, so

$$p_3 = (.9 \times .9 \times .9 \times .407)/6$$
$$= .049 \text{ or } 4.9\%.$$

Thus the theoretical expectation is that there would be three strikes in 4.9 percent of all weeks, or in $0.49 \times 626 = 31$ weeks. This is close to the observed figure of 28 weeks.

The Shape of the Poisson

Poisson distributions can take three different shapes, as illustrated in Figure 5.5. The shape depends on the value of the mean μ:
  (a) For μ less than 1, the distribution is reverse-J-shaped.
  (b) If μ is greater than 1 but less than roughly 4, the distribution is hump-backed but skew with a long tail to the right.
  (c) As μ gets larger than 4, the distribution becomes increasingly symmetrical. It can then be modelled alternatively by a Normal distribution (as a limiting case) with the variance equal to the mean. This makes it easier to

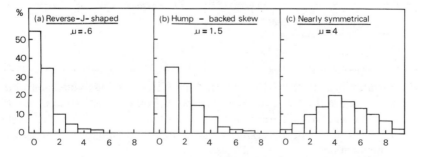

FIGURES 5.5   Poisson Distributions with Various Means

calculate how many readings should lie between any particular pair of values.

Various observed phenomena have been found to follow Poisson distributions to a fair degree of approximation; e.g. the incidence of certain kinds of cosmic radiation in physics, breakdowns in telephone equipment, arrivals of patients at a doctor's clinic, and the numbers of white cells in small samples of blood. However, the simplicity of the Poisson—only one parameter and the variance always equal to the mean—restricts the kind of data to which it can be fitted. There are many types of counts which the Poisson cannot describe. Sometimes certain extensions of the Poisson distribution can provide a more suitable model.

The Negative Binomial Distribution

One possible extension of the Poisson is called the Negative Binomial Distribution (NBD). This consists of a mixture of Poisson distributions with different means. The NBD model makes the same assumptions (a)–(c) given at the beginning of this section for the Poisson: i.e. that the events counted occur irregularly, independently of each other and with no effective upper limit. But the NBD does not assume a constant average rate or probability of occurrence. Instead, the rate or probability can vary from case to case. This greatly extends the model's range of possible applications.

An example is consumers' purchases of frequently-bought branded goods. The top line in Table 5.7 sets out the number of times a sample of households bought Kellogg's Corn Flakes in half a year. The average rate is 3.4 purchases per household and the variance is 27. This difference means that a Poisson with $\mu = 3.4$ would not fit, as the second line of the table shows in detail. The observed distribution is far more skew than the Poisson. However, the bottom line of the table shows that the data can be closely fitted by an NBD.

TABLE 5.7   The Number of Purchases of Corn Flakes in Half-a-Year
(Percentage households buying: Observed, Poisson, and Negative Binomial Distributions)

		Number of Purchases in Half-a-Year										
		0	1	2	3	4	5	6	7	8	9	10+
% buying Corn Flakes	Observed %	39	14	10	6	4	4	3	3	2	2	13
	Poisson %	3	11	19	22	19	13	7	4	1	1	-
	NBD %	35	16	10	7	6	5	3	3	2	2	11

The assumptions of the NBD model imply that an individual consumer's purchases of Corn Flakes over time follow a Poisson distribution, but that different consumers have different average rates of purchase (i.e. that the different Poissons have different means). This is also a realistic assumption in many other practical applications.

The formula for p_r, the proportion of times a count of r occurs, is given in Table 5.8. The calculations are not difficult, but they are tedious and are usually done on a computer. The NBD has two parameters, the mean μ and the 'exponent' k. (The exponent's negative sign in the formula gives rise to the name of the distribution.) The numerical values of μ and k can be calculated from the mean and variance of the observed data, as shown in the table. The variance of an NBD is *greater* than its mean.

TABLE 5.8   The Formula for the Negative Binomial Distribution

The proportion p_r of times a count of r occurs, for $r = 0, 1, 2, \ldots$, is

$$p_r = \frac{(k+r-1)!}{r!\,(k-1)!}\left(\frac{\mu+k}{k}\right)^{-k}\left(\frac{\mu}{\mu+k}\right)^{r}.$$

The two parameters μ and k can be calculated from the mean and variance of the observed data by noting that $\mu = $ mean and $k = \text{mean}^2/(\text{variance}-\text{mean})$.

The NBD can take two shapes. It is reverse-J-shaped if k is less than 1 *or* if μ is less than $k/(k-1)$ when k is greater than 1. Otherwise the distribution is hump-backed but skew. (When the mean is large, the distribution becomes nearly symmetrical.)

Table 5.9 illustrates the usefulness of such a model in describing and comparing four different data sets: purchases of Corn Flakes and of Puffed Wheat over 12 weeks and 24 weeks. The four sets differ markedly, yet they can all be summarized by saying that they are approximately Negative Binomial with specified means and standard deviations (or k-values). From such a summary the full data could be closely reconstructed.

TABLE 5.9   Frequency Distributions of Purchases of Corn Flakes and Puffed Wheat in Different Length Time-Periods
(Percentage of households buying)

	Number of Purchases											Total
	0	1	2	3	4	5	6	7	8	9	10+	
Corn Flakes												
% buying in 24 weeks	39	14	10	6	4	4	3	3	2	2	13	100
" " " 12 "	51	15	8	6	5	5	2	3	1	2	2	100
Puffed Wheat												
% buying in 24 weeks	84	9.6	2.4	1.0	.6	.6	.4	.2	0	.2	1.0	100
" " " 12 "	90	6.3	1.4	.6	0	.4	.2	.2	0	.4	.2	100

5.3   The Binomial Distribution

The ordinary or *positive* Binomial distribution is another model for counting data. Despite its name, this distribution differs from the NBD of the last section. It is usually referred to simply as the Binomial distribution, but sometimes it is called the Bernoulli distribution, after the seventeenth century Swiss mathematician Jacob Bernoulli.

Like the Poisson, the Binomial distribution applies in cases where the events or items being counted occur independently of each other, irregularly, and at a constant average rate or probability. But there are two key differences, namely that the Binomial distribution applies to situations where (i) *the non-occurrence* of the event or item is also being counted, and (ii) there is a fixed upper limit to the things being counted.

The phrase 'bi-nomen' means two names and reflects that one is counting how many occurrences out of a certain total fall into each of two classes, the event and its non-occurrence. In the last section we considered data where the number of strikes was counted over a fixed time-period; the converse—the number of non-strikes was not measured. But if we were to take the total number of trade unions, which is more or less fixed from week to week, and then count the numbers on strike and not on strike in any week, we would have data where the Binomial might apply.

By a simple extension one can apply the ordinary Binomial model to *multinomial* situations, where items or events are classified into more than two categories. Examples are Yes, No, Don't Know responses to questions, or people categorized as single, married, divorced, or widowed.

An Illustration

To illustrate the Binomial distribution, consider the incidence of faulty items in a manufacturing process where the product is packed in batches of twenty units. The top line of Table 5.10 gives some readings for a certain production run of batches of twenty: 11 percent of the batches had no faults, 30 percent had one fault, 25 percent had two, and so on. On average across all the batches, two items per batch of twenty were faulty. That is 10 percent or a proportion $p = .1$.

TABLE 5.10   The Fit of a Binomial Distribution with $p = .1$
(The observed percentage of batches of $n = 20$ items with various numbers of faults)

		Number of Faulty Items per Batch								
		0	1	2	3	4	5	6	7	8-20
% of Cases	Observed %	11	30	25	22	8	3	.4	.6	-
	Binomial %	12	27	29	19	9	3	.9	.2	.05

Faults and non-faults are being counted here, and the upper limit of items is fixed at twenty. A Binomial distribution with a proportion $p = .1$ (i.e. an average rate of 10 percent) should apply to the data if the situation also satisfies all the other assumptions noted above. To test this we calculate the theoretical values of the Binomial. They are shown in the bottom line of Table 5.10: 12 percent of all batches of twenty should have no faults, 27 percent should have one fault, etc.

The agreement between the observed and theoretical figures is generally close: both the observed and theoretical figures say that about 12 percent of batches have no fault, about 30 percent have one, and so on. In such comparisons it is often helpful to group the data. Thus the theoretical distribution says 68 percent of batches should have no to two faults, while the observed distribution showed 66 percent. In the 'tail' of the distribution, the theory says that 1.15 percent should have more than five faults and 1.00 percent were observed. The fit is close.

Where the Binomial distribution gives such a close fit, it provides an efficient summary of the data: 'The incidence of faults is Binomial with an overall proportion of 10 percent.' The fit also tells us something about how the faults arise, i.e. that they appear to occur independently: If one faulty item occurs, it does not mean that succeeding items are more likely to be faulty, which would have implied that the manufacturing process had gone out of adjustment.

The Mathematical Formula

The data for a Binomial distribution must consist of sets of items or events, each set having the same fixed number n. One then counts the proportion p_r of sets having r items with the characteristic in question and $(n - r)$ items without it (e.g. faulty or non-faulty items in sets of $n = 20$). The Binomial formula for the proportion of sets with r items with the characteristic is

$$p_r = \frac{n!}{(n - r)! \, r!} \, p^r \, (1 - p)^{n-r}.$$

Here the symbol p without a suffix traditionally stands for the *overall proportion* of events or items having the characteristic in question. In our example $p = .1$, i.e. the 10 percent of items with faults. As before, the symbol p_r with suffix r stands for the proportion of all sets which take the particular value r. The factorial symbol '!' as in $n!$ again stands for the specified number multiplied by every positive whole number below it.

The Binomial distribution has two parameters: n, the fixed maximum number that can be counted in each set, and p, the overall proportion of events or items having the property in question. (Here n is not to be confused with the symbol n for the total number of readings in a given data set.) It is customary to let the symbol q stand for $1 - p$, the overall proportion of items or events *not* having the characteristic. Thus $p + q = 1$.

Calculating the theoretical Binomial frequencies p_r by the above formula is laborious, especially if n is large. Nowadays computer programs help with the work. But to illustrate the formula, we consider an example with a small value of n, namely the incidence of r boys and $n - r$ girls in families with $n = 5$ children. The data are shown in the top line of Table 5.11 and come from a comprehensive study in 1884 of about 5 million birth registrations in Saxony.

TABLE 5.11   The Distribution of Boys in Families with Five Children

	Number of Boys in Family						Average Proportion of Boys
	Five	Four	Three	Two	One	None	
Observed %	3.7	16.7	32.2	30.5	14.0	2.9	51.4
Binomial %	3.6	17.1	32.1	30.3	14.2	2.7	51.5

The overall proportion of boys was .515 for all family sizes. To assess whether a Binomial Distribution with this general value of $p = .515$ fits the data well, we have to calculate the theoretical values of p_r with $n = 5$ and $p = .515$, as shown in the bottom line of Table 5.11.

The proportion of families with $r = 3$ boys, say, would be given by the following calculations:

$$p_r = \frac{n!}{(n-r)! \, r!} p^r (1-p)^{n-r}, \text{ which for } p = .515, n = 5, \text{ and } r = 3$$

$$= \frac{5!}{2! \, 3!} (.515)^3 (.485)^2$$

$$= \frac{5 \times 4 \times 3 \times 2 \times 1}{(2 \times 1) \times (3 \times 2 \times 1)} \times .137 \times .235$$

$$= .321 \text{ or } 32.1\%.$$

Thus in a Binomial distribution 32.1 percent of five-children families would have three boys and two girls. The fit of this and the other theoretical calculations to the observed data is very close.

The Mean and Standard Deviation

The mean and standard deviation of a Binomial distribution can be calculated in the usual way, but there are also theoretical short-cut formulae.

To illustrate both procedures, the mean of the theoretical Binomial distribution in Table 5.11 can be calculated simply as

$$\text{Mean} = \frac{\text{Sum} (f \times x)}{\text{No. of readings}}$$

$$= \frac{5 \times 3.6 + 4 \times 17.1 + 3 \times 32.1 + 2 \times 30.3 + 1 \times 14.2 + 0 \times 2.7}{100} = 2.575.$$

(The denominator is 100 because the relative frequencies refer to a base of 100.) Thus there are on average 2.575 boys per five-child family.

The *theoretical* formula for the mean of a Binomial is

$$\text{Mean} = np.$$

This only uses the two parameters n and p, here the number of children in each family times the proportion who are boys. For our example this comes to $5 \times .515 = 2.575$, the same as by the direct calculation.

The theoretical formula for the standard deviation of the Binomial distribution is

$$\text{Standard deviation} = \sqrt{\{np(1 - p)\}}.$$

This is often written as $\sqrt{(npq)}$, with q standing for $(1 - p)$. In our example the formula gives

$$\sqrt{\{np(1 - p)\}} = \sqrt{\{5 \times .515 \times .485\}}$$
$$= \sqrt{1.25} = 1.12.$$

This figure agrees with the direct calculation of the standard deviation of the theoretical distribution from the bottom line of Table 5.11:

$$\sqrt{\left\{ \frac{(5 - 2.575)^2 \times 3.6 + (4 - 2.575)^2 \times 17.1 + \ldots (0 - 2.575)^2 \times 2.7}{100} \right\}} = 1.12.$$

The *variance* of a Binomial distribution is $np(1 - p)$ or npq. This is always less than the mean np since $(1 - p)$ is always less than 1 (except when $p = 0$). In our example the mean is $np = 2.757$ and the variance is $1.12^2 = 1.25$.

The Shape of Binomial Distributions

Binomial distributions can take three different shapes depending on the values of n and p:

  (a) The Binomial distribution is reverse-J-shaped when p is less than $1/(n + 1)$. In effect this amounts to the mean np being less than 1. Figure 5.6a illustrates this shape for $n = 100$ and $p = .006$. The distribution is very skew, with most of the readings being zero. When p is greater than $n/(n + 1)$, the distribution is correspondingly J-shaped, with most of the readings at the top end.

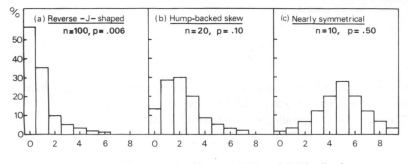

FIGURES 5.6   The Shapes of Binomial Distributions

(b)   The Binomial is hump-backed but usually skew when the mean np is greater than 1. The tail is to the right if p is less than .5, as illustrated in Figure 5.6b. The tail is to the left if p is *greater* than .5.

(c)   The Binomial is approximately symmetrical when p is close to .5 or for other values of p when n is very large. Figure 5.6c illustrates the shape for $p = .5$ and $n = 20$. The closer p is to .5, the more symmetrical the distribution will be even for small or moderate n.

Two special 'limiting cases' are useful to know about because the Binomial then approximates simpler distributions:

(i)   If p is very small (as in Figure 5.6a), then $(1 - p)$ is virtually equal to 1. The Binomial variance $np(1 - p)$ will therefore be almost equal to the mean np, as in a Poisson distribution. In practice, a Poisson distribution with mean $\mu = np$ will then give a good fit and will be easier to calculate. (The proportion of high readings of r will be negligible in a Binomial with a small mean. Hence it does not matter if the theoretical Poisson distribution does not have a fixed upper limit.)

(ii)   When the Binomial distribution is symmetrical the frequencies in any specified interval can be described more simply by a Normal distribution with a mean equal to np and a standard deviation of $\sqrt{\{np(1 - p)\}}$. The degree of approximation is closer the larger the value of n.

There are various extensions of the ordinary Binomial distribution which accommodate a wider range of data. One of these is the Beta-Binomial Distribution (BBD).

The Beta-Binomial Distribution

The BBD relaxes the condition that different data sets all have to have the same average rate or probability of the items in question occurring, e.g. boys in a family.

As an illustration we consider again the incidence of boys in five-children families, as shown in Table 5.11 earlier. The ordinary Binomial with $p = .515$ gave a good fit to the observed data, as it also does to the data on families with other numbers of children.

Nonetheless, the fit is not perfect. In Table 5.11 the observed proportions of all-boy and all-girl families were just fractionally larger than the theoretical values (e.g. 2.9 percent versus 2.7 percent for 'no boys'). The same occurs for other size families. This means that one or more of the assumptions listed earlier does not quite apply. One possible explanation is identical twins, i.e. lack of independence in certain cases, but identical twins are too rare to account for the discrepancies fully.

Another possibility is that the probability of boys is not constant for different families. The Beta-Binomial Distribution (BBD) allows for this. It has three parameters: n, which here stands for the fixed number of events or items in each set, α (Greek 'alpha') which reflects the overall average incidence of the characteristic in question, and β (Greek 'beta') which reflects the variability in the rate of incidence among the different data sets. The values of α and β can be calculated from the mean and variance of the observed data by solving the two equations shown in Table 5.12. This also gives the formula for calculating the theoretical values of p_r, the proportion of cases with r items (e.g. the proportion of families with r boys out of n children).

TABLE 5.12   The Formula for the Beta-Binomial Distribution

The proportion p_r of a count of r out of n is

$$\frac{\alpha!}{(n-r)!\,r!} \; \frac{(\alpha + r - 1)!\,(n + \beta - r - 1)!}{(n + \alpha + \beta - 1)!} \; \frac{(\alpha + \beta - 1)!}{(\alpha - 1)!\,(\beta - 1)!} \; .$$

(The factorials ! should strictly be written as Gamma-functions for non-integral α and β.)

The values of the two parameters α and β can be calculated from the mean and variance of the observed data by solving the equations

$$\text{mean} = n\alpha/(\alpha + \beta) \; ,$$

$$\text{variance} = n\alpha\beta(n + \alpha + \beta)/(\alpha + \beta)^2 (1 + \alpha + \beta) \; .$$

The fit of the BBD to the observed data shown in Table 5.11 is even closer than that of the ordinary Binomial given there. This occurs for families of all sizes. It implies some variation from family to family in the general likelihood of having boys. The size of such variation would however be small, given the close fit already of the ordinary Binomial distribution.

More generally, the Beta-Binomial is a very flexible distribution. It can be reverse-J-shaped, hump-backed, or—unlike the other theoretical models—U-shaped. Table 5.13 shows an example for U-shaped data: the different numbers of issues of a certain weekly magazine bought by UK adults over a 12-week period. The discrepancy for those buying just one issue is fairly large (4.2 percent versus 2.0 percent), but the model closely reflects the general variation in the observed figures.

TABLE 5.13   A U-shaped Distribution and the Fit of the Beta-Binomial

		Number of Issues Bought												
		0	1	2	3	4	5	6	7	8	9	10	11	12
Observed	%	85	4.2	1.7	.8	.4	.3	.2	.4	.3	.5	1.1	1.2	3.8
Beta-Binomial	%	87	2.0	1.1	.8	.7	.6	.5	.6	.6	.6	.7	1.1	3.8

5.4   Discussion

Theoretical models can sometimes provide a simple description of observed frequency distributions. Only the type of model and its parameters need then to be specified.

This process is exceptionally simple with the Normal distribution because it always takes the same basic shape: two-thirds of the readings lie within ±1 standard deviation from the mean, 95 percent within ±2 s.d., and nearly all the readings lie within ±3 s.d.

Skew data are generally harder to describe and the theoretical models are not as simple as the Normal. Indeed, quite often no successful model is available at all.

The fit of a theoretical model to observed data is judged by the size of the discrepancies between the two sets of figures, compared with the variation in the observed readings. The overriding criterion is whether there are *systematic* deviations, i.e. whether any discrepancies occur consistently across different data sets.

Although theoretical distributions are often specified in terms of their means and standard deviations, other ways of designating their characteristic values or parameters can be useful. With reverse-J-shaped distributions, for instance, it may be preferable to state the proportion of zeros and the average of the non-zero readings. For example, in analyzing how long airplanes have to wait before landing, it could be informative to note the percentage of planes which do not have to wait at all and the average waiting time of those which do.

The five types of theoretical distributions discussed in this chapter are summarized in Table 5.14. They cover only a limited range of the models which exist.

Choosing an appropriate theoretical model for observed data is generally not straightforward. There are no cook-book rules. It usually requires trial-and-error, plus hard work and luck. Sometimes a clue comes from conjectures about a

TABLE 5.14 The Five Theoretical Distributions Discussed in this Chapter

Theoretical distribution	Shape	Range	Type of variables	Parameters	Formula
Normal	Humpbacked symmetrical.	Minus infinity to plus infinity.	Continuous*	μ, σ	Section 5.1
Poisson	Reverse-J, OR: Humpbacked skew to the right, OR: Humpbd. near-symmetrical.	0 to infinity.	Discrete	μ	Section 5.2
Negative Binomial (Mixture of Poissons)	Ditto.	Ditto	Ditto	μ, k	Table 5.8
Binomial	Reverse-J or J, OR: Humpbacked skew, OR: Humpbd. symmetrical.	0 to n (fixed upper limit).	Discrete	n, p	Section 5.3
Beta-Binomial (Mixture of Binomials)	Reverse-J or J, OR: Humpbacked skew, OR: Humpbd. symmetrical, OR: U-shaped.	Ditto	Ditto	n, α, β	Table 5.12

* But can also be used for counts

possible underlying process that might have caused the observed data. Such processes are discussed in Chapter 6.

CHAPTER 5 GLOSSARY

Bernoulli distribution	Another name for the Binomial.
Binomial distribution	The ordinary or positive Binomial distribution which can model counts of occurrences that fall into one of two categories.
Beta-Binomial (BBD)	A mixture of Binomial distributions where the average rate or probability of occurrence varies between different sets.
Continuous distribution	A distribution where the variable can take any numerical value within a given range (which may extend infinitely).
Counts	The number of times something occurs (e.g. 0, 1, 2, 3, etc. times).
Discrete distribution	A distribution where the variable only takes selected values, usually whole numbers like counts.
e	A mathematical constant which equals 2.718.
Factorial	The 'factorial' $n!$ means $n(n-1)(n-2) \ldots (3)(2)(1)$.
Gaussian distribution	Another name for the Normal distribution.
Lognormal distribution	A distribution where the logarithms of the measurements follow a Normal distribution.
n	In the Binomial distribution the symbol for the fixed size of each set.
n!	! is the symbol for a factorial (as in $n!$).
Negative Binomial (NBD)	A mixture of Poisson distributions, where the average rate or probability of occurrence varies for different cases.
Non-linear transformation	Changing a scale of measurement so that a unit difference between two large values is changed to a different extent than a unit difference between two small values.
Normal distribution	A formula which describes data that follow a particular hump-backed and symmetrical frequency distribution.
p	The proportion of occurrences in a Binomial distribution which have one of the two possible outcomes.
p_r	The proportion of readings taking the value r.
Parameter	A characteristic value of a theoretical distribution (e.g. the mean, the standard deviation, or the proportion of zeros).
Poisson distribution	A theoretical formula that under certain conditions can model counts of events that have no fixed upper limit.
q	Equals $(1-p)$ in a Binomial distribution, i.e. the proportion of occurrences which have the *other* outcome.
Theoretical distribution	A mathematical formula that describes a particular shape or kind of frequency distribution.

CHAPTER 5 EXERCISES

5.1 In a Normal distribution with mean $\mu = 20$ and standard deviation $\sigma = 4$:

(a) What percentage of readings will be greater than 28? Betweeen 12 and 28? Lower than 12 or greater than 28?

(b) How large is the mean deviation?

(c) Explain how this theoretical distribution can fit data on the heights of a set of seedlings, with mean 20 mm and standard deviation 4 mm, when none of the heights can be negative?

5.2 To what extent are the following three sets of $n = 20$ readings approximately Normally distributed:

Set 1   7, 8, 9, 9, 10, 10, 11, 11, 12, 12, 12, 12, 13, 13, 13, 14, 15, 15, 16, 18. $m = 12$, s.d. $= 2.8$.

Set 2   7, 8, 9, 9, 9, 10, 11, 12, 12, 12, 13, 13, 13, 13, 13, 14, 14, 15, 16, 17. $m = 12$, s.d. $= 2.7$.

Set 3   9, 9, 9, 9, 9, 9, 9, 9, 10, 10, 11, 12, 12, 13, 13, 14, 15, 15, 19, 24. $m = 12$, s.d. $= 4.0$.

5.3 The widths in meters of ten square plots are

$$1, 2, 6, 8, 8, 9, 9, 10, 11, 12.$$

Is the distribution of these widths or that of the plots' *areas* more nearly Normal?

5.4 The following data show what percentage of ships in a fleet had 0, 1, 2, 3, etc. minor accidents in 1980:

No. of accidents	0	1	2	3	4	5	6	7+
Percentage of ships	19	34	26	14	4	2	1	0

(a) What is the average number of accidents per ship? Calculate the variance of the distribution and comment on the implications for fitting a Poisson distribution.

(b) Fit the first six terms of a Poisson distribution to the data. Using $\mu = 1.6$, first work out $p_0 = e^{-\mu}$, then p_r from $p_r = (\mu p_{r-1})/r$. Multiply the theoretical proportions p_r by 100 to compare them with the observed percentages in (a).

5.5 (a) In a study of the distribution of boys and girls in 5000 families, the data were tabulated in two ways:
(i) The numbers of all families who had 0, 1, 2, 3, etc. boys.
(ii) The numbers of families with a given number of children (say five) who had 0, 1, 2, 3, etc. boys.
For each set of data might a Binomial distribution be a possible model?

(b) In Table 5.10, a Binomial with $n = 20$ and $p = .1$ was fitted to data on the number of faulty items. Recalculate the theoretical percentage of batches you would expect to have two faults.

(c) What is the mean and the variance of the theoretical distribution? (Assume that the average number of faults between 8 and 20 is 10.)

5.6 The following table shows a Beta-Binomial distribution fitted to the percentages of adults in Germany who saw from 0 to 11 episodes of a TV serial based on Thomas Mann's *Buddenbrooks*. Comment.

	Number of episodes seen											
	0	1	2	3	4	5	6	7	8	9	10	11
Observed	32.4	12.5	9.4	7.3	4.5	4.2	5.3	4.8	6.0	5.0	6.1	2.8
Theoretical	31.0	12.7	9.0	7.0	6.3	5.6	5.2	4.8	4.6	4.4	4.4	4.6

CHAPTER 6

Probability Models

Probabilities can be used to measure the chance that an *as if random* event has a particular outcome. We say *as if* random because it is impossible to determine whether the outcomes of observed events are truly random—i.e. occurring without *any* systematic order. True randomness is only an abstract concept.

But for practical purposes, if a long sequence of observations shows that a particular type of event has sufficiently irregular outcomes, it is often useful to describe it as behaving in an 'as if random' manner. This allows us to use probabilities to describe the event's outcomes. For example when tossing a well-balanced coin, over a long series of tosses virtually half the outcomes will be heads and half tails, but in no discernible pattern. Thus for each toss of the coin we can say the probabilities will be .5 for heads and .5 for tails.

Probabilities are measured on a scale running from 0 for no probability to 1 for certainty. The values are usually determined either by how often the event's observed outcomes occur in the long run (the 'frequency definition' of probabilities), or by one's 'degree of belief' about the event's likely outcome.

The main attraction of probabilities is that in certain circumstances they allow us to deduce the likelihood of *combinations* of events. Thus on tossing a coin twice, we can calculate the likelihood that it will come down heads both times as $.5 \times .5 = .25$. The mathematical rules for multiplying and adding probabilities are outlined in a technical note at the end of this chapter.

Probabilities have a number of different uses, e.g. to describe frequency distributions, to explain how such distributions might arise, and to judge uncertain events. Probabilities are also used in statistical sampling, as discussed in the next chapter.

6.1 Probability Distributions

One use of probabilities is in describing frequency distributions. In the last chapter we used proportions or percentages, e.g. that 68 percent of the readings in a Normal distribution lie within 1 standard deviation on either side of the mean. But we could also say that any one reading, however selected, has a .68 probability of lying within these limits. Which measurement takes a particular value is then taken to be a matter of chance. When theoretical distributions are expressed in these terms they are called *probability distributions*.

Probabilities are easier to handle mathematically than proportions because a probability is about an individual case: the probability of an individual reading lying between $\pm 1\sigma$ from the mean is .68. In contrast, proportions or percentages are about *groups* of readings: we have to consider all the readings in a set of data to be able to say that 68 percent lie between $\pm 1\sigma$.

For observed data it makes sense to turn frequency distributions into probability distributions if the different outcomes in question appear to occur irregularly. For example, we might have a manufacturing process where faults occur 10 percent of the time, but without any general pattern (e.g. not just mostly at the end of the day). We might then want to say that any item in the process has a .1 probability of being faulty.

At a more complex level there might be systematic patterns in the data (e.g. 2 percent of items faulty in the mornings and 18 percent in the afternoons). But if within each day-part the data still appeared irregular, we could regard the incidence then 'as if' random and use probabilities to describe the faults.

6.2   Probability Processes

Theoretical frequency distributions—like the Normal, Poisson, and Binomial—can sometimes be explained by what is called a *probability process*.

A probability process is a theory about how certain chance factors might combine. Different combinations would result in different probability distributions, as we will describe. Probability processes are sometimes referred to as *stochastic* processes (the word stochastic coming from the Greek 'stochos', meaning a guess or chance).

In cases where a theoretical distribution fits the observed data closely, the underlying probability process can then provide a possible explanation of how the observed data came about. We are familiar with the idea that on tossing an evenly-balanced coin getting heads or tails is like a 50:50 chance process. This 'explains' why we sometimes can get a run of five heads in succession 'by chance', i.e. without having to suppose that the coin has suddenly become biased.

The general argument has four stages:

(a) There is an *observed* frequency distribution.

(b) There is a *theoretical* distribution, i.e. a formula, which provides a close description of the observed data and which can be expressed as a theoretical probability distribution.

(c) There is an *underlying probability process* which gives an exact mathematical reason for the occurrence of the theoretical probability distribution.

(d) This probability process then gives a *possible explanation* of the observed data.

Such an argument does not prove that the observed phenomena really occur by chance. It only shows that they occur 'as if' by chance. But it will also help to rule out other possible explanations for their occurrence. We now briefly describe several kinds of probability processes, starting with the Binomial.

The Binomial Process

Suppose we looked at a large number of families with three children and measured how many of the children were boys. We would get figures looking something like those in Table 6.1.

TABLE 6.1   **The Number of Boys in Families with Three Children**

	Number of Boys in Family				Av. % of Boys
	Three	Two	One	Zero	
% of Families:	13.9	38.3	36.2	11.6	51.5

In 100 families there are 300 children and the average proportion of boys is $(3 \times 13.9 + 2 \times 38.3 + 1 \times 36.2 + 0 \times 11.6)/300 = .515$ or 51.5 per cent.

One possible explanation of these observations is that just over 10 percent of families are only capable of having boys; about another 10 percent can only have girls; about half of the rest have a built-in physiological bias towards boys and the other half have a bias towards girls. That is what the data look like. But an alternative hypothesis is that all families are alike; that there is a 51.5 percent probability of any baby being a boy; and that the number of boys each family actually has is in effect a matter of chance.

The latter mechanism is called a Binomial or Bernoulli process. It results in the Binomial distributions described in Section 5.3. If we have sets of n events, where each event can have one of two outcomes, X or not-X, this probability process supposes that the different frequencies in the distribution arise because of an underlying combination of the following factors:

(a)   Over the long-term outcome X occurs in a steady proportion, p, of the events.

(b)   Each outcome is independent of all the other outcomes.

(c)   Whether a particular event has outcome X is a matter of chance, so outcome X occurs with probability p.

If these three stipulations are met, it can be shown mathematically that the frequency with which outcome X occurs in different groups of fixed size n must follow a Binomial distribution.

The Theoretical Argument

To illustrate the mathematical argument, we consider how many families with $n = 3$ children should have three boys, two boys, one boy, or no boys if the three conditions were precisely met and the overall probability of boys were .515.

First we look at the theoretical probability of a three-child family having all boys. Because the Binomial process assumes that individual outcomes are independent with a fixed probability $p = .515$, the probability of the first child being a boy is $p = .515$, and it is the same for the second and the third child. Under the Multiplication Rule for Independent Probabilities (discussed in the end-of-chapter technical note), if two events are independent, the probability of both having the same outcome equals the product of their separate probabilities. Thus the probability of two babies both being boys is $.515 \times .515 = .265$. Similarly, the probability of three babies all being boys is $.515 \times .515 \times .515 = .137$.

The probability of a girl, outcome not-X, is $(1 - p) = (1 - .515) = .485$. The probability of a family having two boys followed by one girl is therefore $.515 \times .515 \times .485 = .129$. But we must also consider the other possible orders of the births. The probability of having one girl followed by two boys is $.485 \times .515 \times .515 = .129$. And the probability of having a boy–girl–boy sequence is $.515 \times .485 \times .515 = .129$. Thus the total probability of a three-child family having two boys and one girl *in any order* is the sum of the three probabilities:

$$.129 + .129 + .129 = .387,$$

(or $.3858$ if working with figures to one more place of decimals).

The corresponding calculations of the remaining possibilities for a three-child family give a $.363$ probability of having two girls with one boy, and a $.114$ probability of having three girls. (The mathematics behind these calculations is that of permutations and combinations.)

These probabilities for three boys, two boys, etc. are those of the Binomial probability distribution. To express them as percentages we multiply by 100. This gives the theoretical relative frequencies of three-child families with three, two, one, or no boys, as set out in the top line of Table 6.2.

TABLE 6.2   The Theoretical and Observed Percentages

		\multicolumn{4}{c}{Number of Boys in Family}	Av. % of Boys
		Three	Two	One	Zero	
% of Families:	Binomial %	13.7	38.6	36.3	11.4	51.5
	Observed %	13.9	38.3	36.2	11.6	51.5

Explaining the Observed Data

Table 6.2 also compares the theoretical distribution with the observed data from Table 6.1. The fit is clearly close. Since the Binomial process accounts for the *theoretical* distribution, the close fit implies that the process could also account for the *observed* distribution. The data could therefore be explained by babies in three-child families all having a probability of $.515$ of being boys, with the outcomes for different babies being independent.

This still does not mean that the incidence of boys is literally a matter of chance. There may well be specific physiological or biochemical reasons for boys occurring as they do in each specific case. The fit of the data only shows that the outcomes are so irregular that they can be successfully modelled on such an 'as if random' basis. The fit of the Binomial model in this single case does not categorically exclude certain other possible explanations; e.g. that families differ intrinsically in their capacity to produce boys or girls, with about 10 percent being capable of having only boys.

However, the latter hypothesis is increasingly ruled out by the fact that the Binomial model also applies to families of other sizes—like those with five-children shown in Table 5.11 in the last chapter. There the incidence of all-boy or all-girl families was much smaller (only about 3 percent instead of over 10 percent). Yet the Binomial distribution with $p = .515$ again gave a close description. The hypothesis of the Binomial process is therefore the more general and more parsimonious explanation.

There may also be other evidence to support the Binomial assumptions. If no generalizable girl–boy–boy or boy–girl–boy patterns or the like are found in births data other than those closely predicted by the Binomial process itself, this lends credence to the Binomial assumptions. The hypothesis is also supported if the proportion of boys is found to be about 51.5 percent for many different groups of babies: e.g. for long series of births observed in hospital; for births at home; for morning or evening births; for summer and winter babies; for different localities, countries and races; for different points in time; for younger and older parents; for different size families; for first-borns and for later children; and irrespective of the sex of the preceding children. These findings would then be consistent with the view that the probability for any baby, however selected, being a boy is about .515.

But as in all scientific explanations, these findings still provide no *proof* that the sex of babies is really determined by chance, or that it precisely follows a Binomial process.

The BBD Process

The fit of the Binomial distribution to the observed births data in Table 6.2 is not perfect, with too few theoretical all-boy and all-girl families. Although the discrepancies are small, they are consistent for a wide range of other data on births (e.g. see Table 5.11).

There must therefore be other theoretical models for the incidence of boys and girls at birth which should give as good a fit or better. The Beta-Binomial distribution (BBD) of Section 5.3 is one of these.

The BBD has a possible underlying probability process with two of the same assumptions as for the ordinary Binomial: i.e. that the sex of a baby is determined as if by chance and independently of the sex of other babies. But the BBD model allows the probability of boys to vary from family to family instead of all being exactly the same at $p = .515$.

As noted in Chapter 5, use of the BBD model only gives a small improvement in the degree of fit here (since the ordinary Binomial with the constant p-value of .515 already gives such a close one). Hence there need only be a small amount of variation from family to family in their probability of having boys. In particular, there still is no need to suppose that large numbers of parents are capable of producing children of only one sex. These arguments illustrate the small backwards-and-forwards steps by which our understanding of observed phenomena often improves.

The underlying BBD process of varying probabilities comes more markedly into play with the U-shaped distribution of magazine readership in Table 5.13. Here the model supposes that each adult has a fixed probability of buying the magazine, and that whether a person buys any particular issue is a matter of chance. But these probabilities will vary greatly between different people. Some have a very high probability of buying and hence buy all or nearly all the issues. Many have a zero or near-zero probability of buying and hence buy none or almost none of the issues.

The Poisson Process

There are similar probabilistic processes which can give rise to the Poisson and NBD distributions described in Section 5.2 of the last chapter.

If we suppose that a certain type of event were to occur at random over time with a fixed probability and independently of previous occurrences, then it can be shown mathematically that the frequency of occurrence of that event in successive time-periods must follow a Poisson distribution.

In practice it then follows that if a Poisson distribution gives a good fit to the observed data, like the incidence of strikes, the implication is that strikes occur as if at random, with a fixed probability or average rate, and independently of previous strikes. Certain other possible explanations of the observed pattern can therefore be eliminated. Thus one strike *might* have triggered off others, or some common factor—like new government legislation, a large cost-of-living increase, or some political plot—could have caused more than one strike to occur in the same week. In Table 5.5 we saw there were 36 weeks out of 626 in which as many as three or four strikes occurred, which seems a high number. Yet the Poisson process tells us that 36 is close to the number one would expect if strikes occurred independently of each other and as if by chance at an average rate of .9 per week.

The NBD Process

A Negative Binomial distribution (NBD) can arise from similar probabilistic assumptions to the Poisson, namely as a *mixture* of Poisson processes. The only difference is that the probability of the event (say, a purchase of Corn Flakes) can vary between the different cases (e.g. for different households).

Such an underlying process says that each household buys the brand at a constant average rate or probability, but that these probabilities vary from household to household. Precisely in which weeks a particular household buys is 'as if random' (i.e. a Poisson process).

6.3   Games of Chance

Games of chance provide a popular illustration of probability processes. Probabilities can often be used to predict the outcome of tossing a coin, rolling dice, playing cards, etc., because under the right conditions the individual outcomes have been found to be highly irregular—as if random.

But even so-called games of chance are not truly random. When tossing a coin there is no inherent reason why heads or tails should be a chance phenomenon and strictly follow a Binomial process. If the coin is placed heads up on one's hand and tossed gently, so that it turns over just once, tails will show every time.

However, heads or tails cannot be successfully predicted when a coin is tossed hard so that it spins several times, and different numbers of times on different tosses. One can then use probability mathematics to represent the chances of the outcomes, as if they occurred at random. The resulting calculations for various combinations of outcomes will be correct to a close degree of approximation, like the proportion of times one gets five heads in succession. Nonetheless, the process is not 'really' random, so somebody trying to find a pattern in the spins of a roulette wheel is not necessarily chasing a complete will-of-the-wisp.

6.4   The Central Limit Theorem

A probability process which can underlie the Normal distribution is known as the Central Limit Theorem. This is of special importance in statistics because it helps to explain why there are so many cases where a Normal distribution provides a good model. The Theorem states that when observations are made up of irregular elements which are, (i) large in number, (ii) independent of each other, and (iii) additive, the resulting variation will follow a distribution which gets closer and closer to a Normal distribution as the number of elements increases. (As mathematicians say, the distribution will be Normal 'in the limit', when the number of elements reaches infinity.)

This process can apply to errors of measurement. They often arise from many different causes, more or less independently of each other, with each error being added to or subtracted from the measurement result. (In the nineteenth century it was noted that errors of measurement often followed what was termed the 'Normal Curve of Errors' and the name Normal has stuck.)

But the conditions of the Central Limit Theorem do not always apply. For instance, if a laboratory assistant is sleepy after lunch and makes more errors then, his errors that day would not be independent of each other.

Other phenomena can also follow a Normal distribution, with the Central Limit Theorem still providing a possible explanation. An example is people's heights. Heights are probably determined by a large number of different genetic and environmental factors, and these may well be independent and additive. (For example, if the presence of factor X causes people to be about $\frac{1}{2}$ inch taller rather than 1 percent taller, it would be an additive rather than a proportional effect.)

But the effects are not necessarily like that. One's height depends on that of one's parents, and sometimes short fathers marry short mothers and produce short children. Yet the facts show that heights of people of the same sex and age are usually Normally distributed, and this suggests that the various factors may in effect act independently.

6.5   Subjective Probabilities

We often hear or make statements like 'I will probably be late', or 'The likelihood that this book will sell a million copies is 1 in 100', or 'The chance of rain tomorrow is .8'. These are loose *predictive* uses of the ideas of probability and chance.

There may be some empirical evidence for the statements (i.e. evidence based on observations). For example, there has just been a drop in atmospheric pressure and this generally tends to be followed by rain. But the probability quoted is seldom based directly on any statistical analysis of the outcome's observed frequency in the past. The probabilities are therefore called 'subjective' rather than 'objective' or 'frequentist'.

Subjective probabilities are often used in predicting a one-off event where there can be no long-run of such occurrences from which to estimate a numerical value. One way of giving some meaning to such probabilities is to relate them to statements about other events with the same expected probability of occurrence. For example, a probability of .01 for a particular book to sell a million copies could be regarded as being like the 1-in-100 odds of a certain horse winning a race, or the Republican Party gaining a majority in the House of Representatives in the next election. In the long run such statements should turn out to be correct 1 percent of the time if the probability assessments were correct. But it is not clear how to check this in practice.

Decision Analysis

Subjective probabilities are widely used in everyday life. They have also been used in academic studies of decision-making. The idea behind this is that since we have to make decisions under uncertainty, it might help if we expressed our uncertainties about the future as numerical probabilities.

To illustrate, consider a car manufacturer faced with the following three options for next year: he can either develop a new Model X, or a new Model Y, or no new model. He needs to consider the likely profitability of these options to help him decide among them.

The manufacturer believes that sales of the proposed Model X will be between 200 000 and 400 000. He feels that he knows for sure that he can place at least 200 000 with the trade and that he cannot manufacture more than 400 000. He then has to express his beliefs about the different sales levels within these limits in the form of subjective probabilities. He might do it roughly as follows:

Possible sales of X	Subjective probability
Less than 200 000	0
About 200 000	.2
About 300 000	.6
About 400 000	.2
More than 400 000	.0

To assess the likely profit or loss if he were to launch Model X, he must then attach profit estimates to the different sales targets, calculate the likely 'pay-off' of the proposed new model, and compare it with his other investment options. (These options might include doing nothing, with the risk of having no new models in the long-term.)

Suppose that for Model X his accountants calculate that the profit on selling 200 000 units would be about $20 million. The estimated probability of such sales is .2, so the likely profit for this sales level is $4 million (i.e. $20 million × .2 = $4 million).

If the estimated profits of selling 300 000 and 400 000 cars are correspondingly $40 million and $50 million, the overall expected pay-off of $38 million can be arrived at like a weighted average:

Sales of X	Profit	Probability	Likely Profit
200 000	$20 million	.2	$4 million
300 000	$40 million	.6	$24 million
400 000	$50 million	.2	$10 million
Overall expected pay-off of launching Model X:			$38 million

Suppose that the second possible new car, Model Y, is expected to be either a winner or a clear flop early on. The manufacturer thinks it has a .2 probability of making a profit as high as $300 million, and a .8 probability of not being put into full production at all, leaving a loss of $10 million:

Profit of Y	Probability	Likely Profit
$300 million	.2	$60 million
−$10 million	.8	−$8 million
Overall expected pay-off of Model Y:		$52 million

The expected pay-off of Model Y is $52 million, which seems more attractive than the $38 million for Model X.

Such an 'expected pay-off' will, however, only materialize if the manufacturer has to make many such decisions. He would have to be concerned with his average profits over a long series of such cases, because the calculated expected pay-off would not occur in any particular instance, but only *on average*—always assuming that his probability and cost estimates were correct.

This kind of decision-analysis is therefore less appropriate if the manufacturer has to concentrate on the short-term because of lack of capital, cash flow, job security, or whatever. Model Y has a .8 probability of giving a $10 million loss. Unless the manufacturer has enough risk capital to cover such a loss and try again, he may go bankrupt before he hits a winner.

Instead of seeking to maximize his average profit, an alternative decision strategy would be to reduce the risk of an unbearable loss. This could be attempted, for instance, by seeking to minimize the maximum loss for any particular decision-problem. If he adopts this type of strategy, he would opt for Model X, where his assessment is that he cannot lose: he expects to make *some* profit at any of the sales levels which he thinks likely. However, the detailed probabilities required to make all such calculations remain subjective, and hence of uncertain validity.

Prior Probabilities

Subjective probabilities and probabilistic decision analysis are often linked to the *Bayesian* approach in statistics. This presupposes that one has previously decided on a probability—called the prior probability—for a given hypothesis, but that there is new and independent empirical evidence which now has to be taken into account. For example the prior probability might be .6 that car sales will be 300 000, and a new market survey might now have to be considered.

The mechanism for calculating a new probability is *Bayes' Theorem*. This says that the new or 'posterior' probability of the hypothesis will depend on an adjustment to its prior probability. The adjustment reflects how likely the new empirical result would be if the original hypothesis were in fact true. If the new data is in line with the original hypothesis (i.e. highly likely), it will increase the probability which we are prepared to attach to that hypothesis.

The mathematical correctness of Bayes' theorem is not generally questioned.

But there remain doubts about the subjective nature of the prior probability estimates and their validity. In each instance there may also be doubt whether the new evidence is independent of the evidence on which the prior probability itself was based. Without independence, Bayes' theorem cannot be applied.

6.6   Discussion

The concept of probability is not easy to understand precisely, although we all have some idea of what it means. One difficulty is that probabilities are used in quite different situations: to make predictions about uncertain events in the future (often through subjective probabilities), or for describing frequency distributions, or for *explaining* such distributions.

Three distinct features are involved when probabilities are used to explain observed data. One is the observed distribution. The second is a theoretical distribution which can be used to describe or model the observed distribution. The third is a probability process (i.e. a set of assumptions) which logically leads to the theoretical distribution and might therefore provide a possible explanation for the observed data:

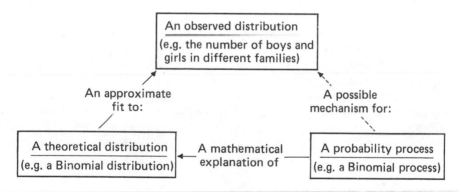

Probabilities are also used in statistical sampling, where chance is deliberately introduced into the data by selecting the sample at random. This will be discussed in Part Three.

CHAPTER 6   TECHNICAL NOTE

The Multiplication and Addition Rules for Probabilities

Two basic rules cover the calculation of probabilities of certain combinations of events. These are briefly summarized here.

The Multiplication Rule for Independent Events

Two measurements or events are *independent* if the outcome of one is not affected by the outcome of the other.

With two independent events, the probability that one will have outcome A and the other will have outcome B is the product of the separate probabilities of A and B occurring, p_A and p_B:

$$\text{Probability of both A and B occurring} = p_A \times p_B.$$

Thus if $p_A = .6$ and $p_B = .3$ and the two events are independent, the probability that both A and B occur is .18. The probability of neither A nor B occurring is similarly $(1 - p_A) \times (1 - p_B) = (1 - .6)(1 - .3) = .28$.

The Addition Rule for Mutually Exclusive Events

If two outcomes X and Y are mutually exclusive, the occurrence of one makes the other impossible. The probability of both X and Y occurring is zero. The probability of either X *or* Y occurring is the sum of their separate probabilities p_X and p_Y:

$$\text{Probability of either X } or \text{ Y} = p_X + p_Y.$$

Thus the probability of a baby being a boy *or* a girl is 1, i.e. the sum of the probabilities of its being either. The probability of a baby being both a girl and a boy is clearly zero.

If the two events are not mutually exclusive, the addition rule in its simplest form does not apply. Instead, the probability of either A or B occurring is the sum of the probabilities p_A and p_B of each happening *minus* the probability that both happen (since both happening is counted twice in adding p_A and p_B). If, in addition, the two outcomes A and B are independent of each other, then combining the addition and multiplication rules means that the probability of either A or B comes out as

$$p_A + p_B - p_A \times p_B.$$

Thus p_A covers the cases where A happens, which means either A only or A plus B. Similarly, p_B covers only B happening, and B plus A. But the probability of *either* A *or* B means adding the probabilities of A only, B only, and of A plus B.

Conditional Probabilities

If two events are not independent but related, the probability of the outcome of one event varies with, or 'is conditional on', the outcome of the other. For example, if N always follows M, the probability of M and N both occurring is simply the probability of M, p_M. In another example, outcome R is three times as likely if outcome S has occurred than if not:

> Probability of R if S has occurred   = .6
>
> Probability of R if S has not occurred = .2

Conditional probabilities often let us deal with more complex situations.

Probabilities as Frequencies

The rules of probability can be illustrated in numerical terms. This requires a slight over-simplification but can help in understanding the rules.

Suppose that we have one hundred people and that the probabilities are reflected by exact proportions, e.g. that a probability of .6 of having dark hair means that exactly sixty out of the one hundred have dark hair. (In practice, the proportions would be exact only if the total number of people were very large, but it is easier to think in terms of 100.)

Suppose the probability of having grey eyes is .3, so that thirty of the people have grey eyes. If the incidence of dark hair and grey eyes were independent, 30 percent of the sixty people with dark hair will have grey eyes. This is eighteen people out of the one hundred, a proportion of .18. It reflects the Multiplication Rule $.6 \times .3 = .18$.

Similarly, 30 percent of the forty with *light* hair will have grey eyes (i.e. twelve). The total number with grey eyes will therefore be $18 + 12 = 30$, or 30 percent of the hundred. This checks with the original probability of .3 for grey eyes.

For two *mutually exclusive* outcomes X and Y, if $p_X = .4$ and $p_Y = .3$, then forty observations out of the hundred will be X and thirty of the hundred will be Y. Since X and Y are mutually exclusive, none of these are the same. The total proportion of X *or* Y is therefore $40 + 30 = 70$, i.e. a proportion .7 out of 100. This is the Addition Rule, that the probability of X *or* Y is $p_X + p_Y = .4 + .3 = .7$.

CHAPTER 6   GLOSSARY

Addition rule	The probability that one or other of two mutually exclusive events will occur is the sum of their separate probabilities.
As if random	Observations so irregular that they appear to occur at random.
Bayesian approach	The attempt to use *Bayes' Theorem* to adjust a prior probability in the light of new evidence.
Bayes' Theorem	A mathematical theorem saying that the final (or posterior) probability of a hypothesis equals its prior probability adjusted in the light of new, independent evidence.
Bernoulli process	Another name for the Binomial process.
Beta-Binomial process	A mixture of Binomial processes with different probabilities. This gives rise to the Beta-Binomial distribution of Chapter 5.

Binomial process   A probability process which assumes that the two possible outcomes, X and not-X, of a certain type of event occur randomly, with a fixed probability, and independently of previous outcomes. If these conditions are satisfied, the frequencies of X and not-X in sets of n events will then follow a Binomial distribution.

Central Limit Theorem   Shows that a Normal distribution arises for observations which are each made up of irregular factors that are large in number, additive, and independent.

Conditional probability   The probability of an event depending on the outcome of another event.

Degree of belief   See Subjective probability.

Decision analysis   An attempt to formalize a theory of decision-making by using subjective probabilities.

Frequency definition   Defining a probability as the relative frequency of the occurrence of an event in the long run.

Games of chance   Games with cards, dice, etc., where the outcome is made so irregular that it appears to occur at random.

Independent events   Events where the outcomes are not affected by each other.

Multiplication rule   The probability that two independent events will both occur is the product of their separate probabilities.

Mutually exclusive events Events where the occurrence of one makes it impossible for the others to occur.

Negative Binomial process A mixture of Poisson processes with different probabilities. This gives rise to the Negative Binomial distribution.

Prior probabilities   The probability of a hypothesis before new evidence is taken into account.

Poisson process   A probability process which assumes that a certain observation X occurs randomly, with a fixed probability, and independently of previous occurrences. The frequency with which X occurs (e.g. in successive time-periods of the same length) will then follow a Poisson distribution.

Probability   A measure of the likelihood of the different outcomes of an event. The concept is used in a variety of situations.

Probability distribution   A theoretical distribution described in terms of the probability of each value occurring.

Probability process   A mathematical formulation which supposes that the outcomes of certain events are due to specified combinations of chance factors. One can then calculate the theoretical frequencies of the outcomes.

Randomness   A theoretical concept that outcomes occur with a definite probability but which outcome occurs in any one case is a matter of chance.

Stochastic process   Another term for probability process.

Subjective probability   A probability value which reflects one's 'degree of belief' that a certain outcome will occur, without direct evidence of the frequency of the outcome in past data.

CHAPTER 6 EXERCISES

6.1 (a) If 20 percent of adults watch the LMN program on television at home between 8 and 8.30 pm on Mondays, might this mean that any adult, however selected, has a .2 probability of watching this program?

   (b) A man visits two friends during the program and expects that the probability that both are watching LMN would be only .04 (i.e. .2 × .2). Comment.

6.2 The probability of a new-born baby being a boy is about .51. This is independent of the sex of other children in the family.

   (a) What is the probability of families with two children having, (i) two boys, (ii) two girls, (iii) one boy and one girl?

   (b) How do you interpret the fact that these three probabilities add to 1.00?

   (c) If a two-child family has a third child, what is the probability of its then having, (i) three boys, (ii) three girls, (iii) one boy and two girls?

6.3 (a) A manufacturing process produces batches of $n = 50$ items. Five percent tend to be faulty, with no systematic trends over time. Which theoretical distribution might you try to describe the data? What other assumptions would be involved?

   (b) What do you conclude if the theoretical model in question gives a good fit?

   (c) If different machines are known to produce consistently different levels of faults, what kind of theoretical model might you try to describe the data?

6.4 The number of planes arriving per hour at an airport are counted. The Poisson distribution does not give a good fit. Comment on possible reasons. Suggest other probabilistic mechanisms which might apply.

6.5 In driving from P to Q, delays occur. On a large number of trips spread across the day, would you expect the distribution of driving times to be approximately Normal? If not, what form might the distribution take?

6.6 Compare the basis for making the following statements:

   (a) 'The probability of a new-born baby being a boy is .51.'
   (b) 'The probability of rain tomorrow is 50:50.'
   (c) 'The probability that Homer was blind is .8.'

PART THREE: SAMPLING

A sample is a selection from a given group or 'population' of items or people. The aim of taking a sample is to save time, effort, and money. The savings are often enormous—e.g. measuring a few hundred people rather than millions.

To be useful, a sample must be fairly typical of the population from which it was selected. Thus if 20 per cent of pregnant mothers at a certain hospital have Symptom X, a good sample of one hundred should have about twenty with Symptom X.

There are various ways of selecting a sample:

$$\boxed{\text{POPULATION}} \rightarrow \begin{array}{c} \text{Selection} \\ \text{procedure} \end{array} \rightarrow \boxed{\text{SAMPLE}}$$

Chapter 7 introduces some of these procedures. A major contribution is the theory and practice of *random sampling*. This selection method ensures that one knows how well the sample is likely to represent the population.

A basic concept in sampling is the *sampling distribution*. This is discussed in Chapter 8. It describes how the results of different random samples would vary from each other.

Reasoning from the observed sample results back to the unknown population figures is called *statistical inference*:

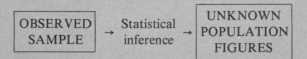

$$\boxed{\begin{array}{c} \text{OBSERVED} \\ \text{SAMPLE} \end{array}} \rightarrow \begin{array}{c} \text{Statistical} \\ \text{inference} \end{array} \rightarrow \boxed{\begin{array}{c} \text{UNKNOWN} \\ \text{POPULATION} \\ \text{FIGURES} \end{array}}$$

There are two main tasks here: (i) Estimating the population values and the likely errors of the estimates (Chapter 9); (ii) Testing a prior hypothesis against the sample results (Chapter 10).

PART THREE: SAMPLING

CHAPTER 7

Taking a Sample

When we want to find out whether a cup of tea is too hot to drink, we usually take just a sip. To take a blood sample, a doctor selects a vein near the surface. In starting up a barrel of beer, the barman runs off a couple of pints and checks the *third* pint to see whether the beer is still cloudy.

None of these samples are statistically representative. Yet the information remains valid because we know how the different parts of the system are related or that the material is homogeneous, so it does not matter which sample item is selected.

Statistical sampling is only needed when there is large and irregular variation in the population and one does not have the time, money, or means to measure the whole population. There is always some loss in accuracy when we use a sample result, but this is acceptable if one knows the likely limits of error.

A good sample of a few thousand is accurate enough for most purposes, and one can often get by with samples much smaller than this. But not just any sub-set of the population will do. The sample should closely represent the population that is being studied.

There are many ways of selecting samples which give systematically biased results, i.e. results which consistently misrepresent parts of the population. This can generally be avoided by using a random or probabilistic selection procedure, as we now discuss.

7.1 Simple Random Sampling

Suppose we wanted to establish how many homes in a certain town (the 'population') had a freezer. Instead of checking on every home, we could take a sample. If we put slips identifying each home into a hat, shuffled them well, and blindly selected the number needed for the sample, we will get a random sample and can go and check on just *their* freezers. Systematic selection bias would be avoided because each home would be chosen by chance.

If we took many such samples, the number of homes having freezers would generally balance out, so that on average we would get the correct population proportion. But in practice we only take one sample, not many. This could be rather unrepresentative. We could be unlucky and get nothing but homes with freezers! With random sampling, however, the chances of getting such an unrepresentative sample can be made small by making the sample size large enough.

To see how this works, suppose 60 percent of the homes in the town have a freezer. (In practice we would not know this beforehand.) If we consider random samples of just three homes, the first home would have a freezer in 60 percent of the samples because it was selected at random and could be any home in the town. The first home therefore has a .6 probability of having a freezer.

The second home in each sample could also be any home in the town. It too has a .6 probability of having a freezer. The same holds for the third home.

Each home was selected independently of the others (i.e. at random). By the multiplication rule (Chapter 6) the probability of all three homes in a sample having a freezer is $.6 \times .6 \times .6 = .216$. Therefore 21.6 percent or just over one in five samples would give a result as dramatically wrong as all homes having a freezer. But a sample size of three is very small. With random sampling we can reduce the chances of such errors by increasing the sample size.

For a random sample of size n, the chance of all n homes having freezers is $.6^n$. If $n = 10$, this is $.6^{10}$ or $.6 \times .6 \times .6 \ldots$ 10 times. This works out to .006, which is less than one in a hundred. For a sample size of thirty, the chance is $.6^{30}$, which is as small as one in a million. Only one in a million random samples of $n = 30$ would be so unrepresentative as to give all 30 homes with a freezer.

Similar calculations apply to the chances of a random sample being atypical in any other respect. (The calculations are those of a Binomial distribution as described in Chapter 5.) The proportion of samples giving results different from the true population value generally decreases for larger samples.

Random samples therefore have three basic properties:
   (i) There is no systematic selection bias. (In certain more complex cases random sampling can involve a *known* bias, but this can then be allowed for in one's calculations.)
  (ii) The chances of obtaining a sample which gives an unrepresentative result can be calculated.
 (iii) These chances decrease as the size of the sample is increased. They can therefore be made small.

Generally the size of the population hardly affects the accuracy of the sample. Only the size of the sample really matters, unless the population itself is small. A sample of 1000 from a population of 50 million is just about as accurate as a sample of 1000 from a population of 50 000.

Selecting a Random Sample

The principles of selecting a random sample are broadly as follows:
  (a) Identify (e.g. list) all members of the population to be sampled. (This gives what is called the *sampling frame*.)

(b) Number each item.

(c) Put each number on a separate slip of paper.

(d) Put these slips into a hat and shuffle them thoroughly.

(e) Blindly select the required number of slips.

This would identify the sample of the population who are now to be measured, e.g. to learn whether they have a freezer, or have Symptom X, or are in favor of the President.

The crucial step is the shuffling. This aims to produce an irregular or effectively random mixture, giving all the slips the same probability of being selected, no matter how one picks them. But since complete irregularity or randomness by shuffling is hard to achieve in practice, we usually add an additional stage to the shuffling by selecting the slips irregularly (e.g. not all from the top).

But selecting large numbers of slips from a hat is hard work. Most random samples are in practice selected from tables of random numbers, as illustrated in Table 7.1. (See also Table A6 in Appendix A.) Such tables are prepared from processes where the results appear irregular enough to approximate theoretical randomness (e.g. certain electronic phenomena). The numbers are then also checked to ensure they show no identifiable sub-patterns.

TABLE 7.1   A Small Extract of 'Random Numbers'

41	03	59	24	78	54	14	48	27	05
53	26	08	33	10	98	62	46	16	94
96	17	25	92	41	17	55	13	73	59
43	61	20	39	65	62	18	15	70	66
65	04	96	78	37	13	98	90	62	28

When using published tables of random numbers, one can start at any haphazardly chosen point (e.g. picking a page and starting-point blindly). One then reads off successive numbers in a predetermined direction, usually horizontally from left to right. To illustrate, suppose we want to select a sample of 5 items from a list of 400 which are identified by three-digit codes going from 020 to 419. We might start with the 08 in the third column and second row of Table 7.1, and then read off groups of three digits, going from left to right. This gives

$$083 \quad 310 \quad (986) \quad 246 \quad 169 \quad (496) \quad 172,$$

identifying a random sample of five. The numbers in brackets are simply ignored because there is no corresponding code in the population.

7.2   Cutting the Cost

Refinements are often used to cut either the cost for a given sample size or the likely size of the sampling error for a given budget. Here we outline three methods used when sampling human populations: multi-stage sampling, clustering, and stratified sampling.

Multi-stage Sampling

When sampling the population of a country or region, it is common to select first a sample of towns or districts, and then individuals in the chosen towns. This is called *two-stage sampling*: first towns or districts, then individuals, as outlined in Figure 7.1.

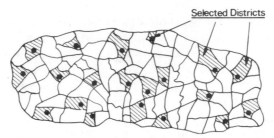

FIGURE 7.1   A Two-stage Sample of Twenty
(First twenty districts, then one individual in each
of these)

    This method avoids the costs of handling lists of individuals in districts where no one is ultimately selected. Instead, only a comprehensive list of the districts is needed, and then lists of all the individuals in the selected towns. *Five* stages would be involved if one selected, (i) towns, (ii) smaller districts in these, (iii) streets, (iv) addresses, and (v) an individual at each address.

Clustering

Instead of selecting twenty towns and one individual per town for a two-stage sample of twenty, one could select ten towns and two individuals from each, as illustrated in Figure 7.2. In surveys involving personal interviews, selecting more than one individual per district cuts traveling costs. (An efficient design for a sample of 900 might be fifteen individuals in each of sixty districts.) Sometimes the term 'cluster sampling' is restricted to cases where every unit at the last stage of multi-stage sampling is selected; e.g. selecting a sample of households and then *all* individuals in the selected households.

    Ideally, clusters should be chosen so that people in each cluster are as diverse as possible. This would reduce sampling errors. In practice the reverse often happens, since people in the same district, street or household tend to be more similar than people in different ones. Thus clustering tends to be less efficient statistically, giving larger sampling errors for a given sample size. The method's cost advantages have to be balanced against the estimated loss of accuracy, although there is often no precise information on the latter.

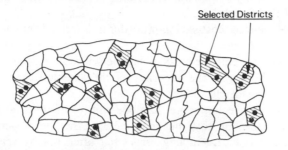

FIGURE 7.2   Clustering
(Ten districts, then two individuals per district)

Stratified Sampling

The statistical accuracy of a sample can be improved at little or no extra cost if the population is first divided into different 'strata' and an appropriate number of people is sampled at random from each stratum. For example, suppose we wanted to select sixty children from a population where boys and girls were 50:50. Few of the possible random samples of sixty one could take would consist of exactly thirty boys and thirty girls. But two samples of precisely thirty boys and thirty girls could be randomly selected, so that the proportions would be exact. The accuracy of measuring any factor related to sex would also tend to be improved.

Often stratification is applied to regions of a country, as illustrated in Figure 7.3. If 22 percent of the population live in Region A, then exactly 22 percent of the sample can be selected from this region.

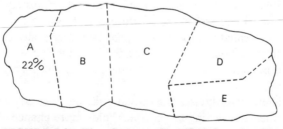

**FIGURE 7.3   The Country Stratified into Its Five
Regions, A to E**

Stratification can be applied at any of the stages in a multi-stage sample design. To do this prior information on the appropriate split of the population is needed. It must also be possible to assign each member of the population to his stratum before the sample is selected. Stratification only improves accuracy if the strata are more homogeneous with respect to the variable being measured than the population as a whole. Many stratifications are used in practice which hardly improve the accuracy of the sample.

To summarize, the three cost-saving procedures described here have the following properties:

Multi-stage sampling cuts the costs of sample selection, but does not affect the cost of data collection or the accuracy of the sample (except when it also involves some form of clustering).

Clustering involves selecting more than one unit (or all units) at a given stage. It reduces data collection costs but may increase the size of sampling errors.

Stratification can reduce sampling errors but often not by very much.

These procedures can be used with any form of sampling, random or other. If random sampling is used without any multi-stage selection, clustering or stratification, it is referred to as *simple* random sampling.

7.3   Probability Sampling

In *probability sampling* items are generally given unequal but known probabilities of selection. This can have advantages over random sampling (in which the probabilities of selection are equal).

Suppose we were selecting individuals from a sample of households. If one individual were chosen from each household, people from large households would be under-represented. This would occur to a known degree if the size of each household were recorded. Then individuals would be selected with unequal but known probabilities. One could redress the imbalance by weighting. Before analyzing the data the results for each sampled individual would be multiplied by the number of people in that household. This is one form of probability sampling.

As an alternative, in each selected household all individuals could be measured, rather than just one. (This would be a form of cluster sampling.) The individuals would then be represented in their correct proportions. An advantage here is that individuals in the same household could be compared if that were of interest. But much of the information in each 'cluster' would probably be highly correlated, and hence redundant.

As a third alternative, households could first be selected with probabilities proportional to their size, giving larger households more chance of being selected than smaller ones. Then one individual could be randomly selected per household. This kind of probability sampling avoids subsequent weighting. It is a form of sampling also used in multi-stage cluster sampling. For example, towns are generally of very different sizes. They can be sampled with probabilities proportional to their sizes, and then an equal number of individuals can be selected in each town.

Which of these forms of probability sampling is used depends on practical factors like costs and accuracy and on the purpose of the study, e.g. whether individuals in a household need to be compared.

Weighting

As we have just seen, in some forms of probability sampling the individual results have to be multiplied by certain factors or *weights* to give them their due importance in the total sample.

Weighting is also sometimes used to reduce the effects of sampling errors. If an unstratified sample of sixty children has resulted in thirty-five boys and twenty-five girls when we know that boys and girls are 50:50 in the population, the results for each sex can be weighted to bring the sample into line with the population proportion. For example, all the girls' readings could be multiplied by $35/25 = 1.4$.

This kind of *posterior stratification* is usually less effective than prior stratification, i.e. getting thirty of each sex in the sample in the first place. The weighting avoids biased proportions in the results, but the more heavily weighted individuals in the sample have an undue effect on the overall results and hence on any random sampling errors. In our example, readings for each girl in the sample would count for 40 percent more than those for each boy (but the weighted readings would correctly reflect the proportions of boys and girls in the population).

Weighting is also widely used when a sample is out of proportion because of 'non-response'. A sample of adults may be short of the very young and the very old, who for different reasons tend to be difficult to interview. Corrective weighting will give the right proportions of old and young people, but not necessarily the right *kinds* of old or young people.

Variable Sampling Fractions

Instead of sampling the same proportion of readings from each stratum or cluster of a population, the proportions can be deliberately varied. A higher proportion may be taken from strata which are of particular interest, or from those in which the variation is known to be greater. This is a common form of probability sampling.

For example, there may be far more patients suffering from a relatively mild attack of a certain illness than from a severe one. In a medical study one might then sample a much higher proportion of those with severe attacks to achieve an adequate sample size for this group. Or in surveying industrial firms, one might select all the large firms, but only a proportion of the small ones.

Before the numbers from the different strata can be combined they have to be brought back into line by weighting. With variable sampling fractions the likely size of sampling errors for the total sample will often be greater than that of a simple random sample of the same size. This has to be balanced against the increased statistical accuracy for some of the sub-groups.

Area Sampling

Often there is no list of all the individuals in the population. The effort of constructing a complete list can sometimes be avoided by using a form of multi-stage probability sampling, so that one need only list a small proportion of individuals. *Area sampling* is one way of doing this, as a form of cluster sampling.

The population is first divided into convenient units, e.g. by geographical areas like city blocks. A random sample of such areas is then selected, as in general multi-stage sampling. Next, the individuals in these chosen areas are listed, as a basis for selecting a random sample of them. A fixed number of individuals can be selected from each area and then the measurements have to be weighted by the size of the area. Alternatively, a fixed *proportion* of individuals can be sampled, the results then being self-weighting.

A form of so-called *random walk* sampling can also be devised when there is no suitable list. This aims to select samples 'on the ground'. For instance, interviewers in a survey might be instructed to use certain specific rules for selecting a suitable sequence of houses or apartments in the designated area ('Walk twelve houses to the left, but turn right if you come to a cross-roads,' etc.). Explicit listing of all the other houses would then not be needed. But unless a random starting-point has been selected, this form of sampling is not really random.

Sampling with Replacement

In the normal process of selecting a simple random sample of size n from a population of N members, N numbered slips are effectively put into a hat, shuffled, and then n slips are successively selected.

With the first selection each slip has a $1/N$ chance of being selected. There are now $(N-1)$ slips left in the hat, so each of the remaining slips now has a $1/(N-1)$ chance of being selected. This continues with chances of $1/(N-2)$, $1/(N-3)$, etc. as successive slips are drawn. This kind of '*sampling without replacement*' is therefore, strictly speaking, a form of probability sampling: items are selected with different but known probabilities. This selection method should be allowed for in the analysis, but would make this relatively complex. (Each successive reading would have to be weighted by factors proportional to N, $(N-1)$, $(N-2)$, etc. before combining and analyzing them all.)

Alternatively, each sampled slip can be replaced and the population of slips reshuffled before the next slip is selected. Then the probability of any slip being selected remains $1/N$ at each stage. This is called '*sampling with replacement*'. Since all the probabilities of selection are equal, the analysis of the results is much easier than for sampling without replacement.

In practice much sampling is carried out without replacement, but for simplicity this is ignored in the analysis. Such fudging is possible because the numerical results of the two types of sampling are almost the same whenever n, the size of the sample, is small compared with N, the size of the population.

Randomized Experiments

Random or probability sampling plays a major role in statistically controlled experiments. Suppose we wanted to assess the effect of an agricultural fertilizer. Experimental control can be introduced by taking some plots of ground and dividing them into two sub-groups *at random*. One sub-group would be treated and the other left alone as a control.

Random selection is important if the material under examination is very uneven. Some plots of ground could be more fertile than others, greatly affecting the results of the study. But in a randomized experiment the chance of the control group having plots with very different fertility than the treated group would be relatively small and could be calculated.

Stratification can improve the statistical accuracy of randomized experiments. For example, all the plots could first be matched according to some available measure of fertility—like last year's yield, the condition of the soil, or geographical location. The plots within each such stratum would then be randomly allocated to the control or experimental groups. The design of statistical experiments is discussed further in Chapter 21.

7.4 The Population Sampled

To be able to select a sample, a good definition of the population is needed (e.g. male adults under 65 living in private households in the UK in March 1981). Otherwise it is not clear what the sample is supposed to represent. One also needs a comprehensive list of all members of that population. This is called a sampling frame.

Preparing such a list is a major task—usually far bigger than the job of taking a sample from it. For instance, suppose we wanted to assess the incidence of Symptom X among children aged under 5 who had had German Measles. To select a sample of 300 children for that survey, we would in principle first have to have a list of *all* such children in the district where the study would be made.

An old list from last year, even if it existed, could not be used because it would be out-of-date. Newcomers to the district would be missing, including children under 1, and all recent measles cases. One possibility might be to construct a sampling frame from health records. But how accurate, up-to-date, and accessible are they?

Another possibility would be to survey all households, children's homes, hospitals, etc. in the area, listing all children under 5. But that would be vastly expensive. To keep costs down, one would like to establish simultaneously whether each child listed had had German Measles. But it is highly unlikely that one's sources of information for children's names would also know their medical history. So a further stage of work would be necessary, surveying all the listed children to try to identify those who had had the illness.

Only when all of that work had been finished would we be ready to select our sample of 300 children from the list and start work on the survey itself, i.e. finding out how many of them had Symptom X.

Thus the technical and cost implications of making a sampling frame can be very large if an appropriate one does not already exist. That explains why many studies do not use formal methods of sampling or carefully defined populations.

Non-Sampling Errors

Often an item or person selected for a sample cannot be measured. Such failures are called *non-sampling errors* because they would occur even if the whole population had been measured. For example, in a biological experiment some plants may die because the experimenter or his staff stepped on them. Or in measuring trees in a forest, some may be inaccessible at a reasonable cost.

In surveys of human populations such problems are referred to as *non-response*. This can be caused by people who have moved, people refusing to be interviewed, or people who are not available for interview at the time of the call.

Non-response will make the sample results unrepresentative of the defined population to the extent that those not covered differ in relevant ways from the people who respond. However, these differences are usually not known.

There are three ways to deal with non-response:

  (i) Try to reduce it, e.g. by avoiding difficult survey questions or by making repeated calls on those initially not available.

 (ii) Redefine the population by treating the sample actually achieved as representing only those who would cooperate in the survey.

(iii) Establish in what relevant respects the non-responses are likely to differ from the respondents. This can sometimes be done by noting easy-to-measure characteristics of the non-responses (e.g. sex, approximate age, working or non-working, social class, etc.). Sometimes special studies of non-responders can provide specific information about them. They tend for example to be either very old or young, and light users of the product or service being studied.

Biased Populations

For technical or cost reasons one may measure a population or a sample from it which is not the exact population or sample wanted. For example, a survey may only cover the 80 percent of the population who normally cooperate; or a biological experiment may not cover sensitive plants which tend to die before such studies are completed. The collected data can still be informative as long as one does not pretend that they are more widely representative than they are.

An extreme example of a biased population is a food manufacturer who tests the acceptability of a new product by using his staff as guinea pigs. This is convenient and inexpensive, but the staff are likely to be far more experienced with such products than is the population as a whole, and are likely to be biased (i.e.

systematically different) in other respects. They certainly are not a statistical sample. But as long as the manufacturer is fully aware of this, such in-house tests may be useful as a first step in his acceptability studies.

If there is no sampling frame for the population of interest (or no *accessible* frame), people often adopt some *ad hoc* selection procedure without explicitly defining the population in question. For instance, if motorists are the people of interest, one might interview every fifth customer at a gasoline station. But such a sample would not be representative of the population of motorists as a whole. What is the population of interest, e.g. 'owners', or 'main drivers', or 'users'? The *size* of this population would also not be known—i.e. how many people in the country use gas stations in a week? Hence one cannot calculate results for the total population (e.g. the size of the gasoline market). The selection procedure would in any case over-represent heavy drivers and under-represent infrequent ones. (To spot biases in a selection procedure, imagine that one is interviewing *all* the people in question, e.g. every customer at all gasoline stations, for a week say. Then heavy motorists would be seen more than once and light motorists only once or not at all.)

Sample results from such ill-defined populations can be very misleading if they are assumed to be representative. But occasionally they can give one a feel for the subject-matter, or help counter some prior hypothesis. For instance, interviews with people leaving accessory shops would not give a representative result for the population of car-owners. But suppose it were thought that a new electronic gasoline-consumption meter would first be taken up by gadget freaks and do-it-yourself motorists. Then it could be helpful to learn that almost none of the people leaving the accessory shops had said they would buy one.

Is There a Population?

It is unfortunately quite a common practice to try to upgrade some haphazard collection of readings by calling it a sample. One often hears or reads 'A random sample of sixty patients was tested.' But what was the population from which they were selected? Often one is not even told what kinds of patients there were in the sample, let alone in the population. What was the nature, duration and severity of their illnesses? Were they adults or children? Without knowing the population, one cannot know what the sample might represent.

Was it really a sample? If the population is not described, probably no explicit sampling took place. The 'sixty patients' may merely be everybody who happened to be examined at a given time. They would then make up a small, possibly ill-defined, population of their own. They would not directly represent any larger group.

7.5   Other Types of Sampling

Because of the frequent difficulties of compiling a valid sampling frame of the population of interest, and because data collection based on random probability samples can itself be expensive (e.g. in sample surveys), many other ways of selecting samples are practiced. Some are more 'quick and dirty' than others. They all aim to cut the cost and bother of the more rigorous procedures. The price paid is that the selection procedure may be biased, yielding samples which consistently fail to represent the population. Furthermore, one cannot usually calculate the likely size of any errors due to measuring only a sample.

Convenience Sampling

Convenience sampling is one popular form of selecting a sample. For example, the first items on a list are selected because that is easiest. ('Let's check the first hundred.') Here it may not be at all clear what the sample tells about the rest of the list. The list may be in some specific order (e.g. alphabetical, or by age, or by date of purchase of some product). Selecting the first hundred could, therefore, lead to consistent bias. Every time such a sample was taken, the same kind of error would tend to be made.

Systematic Sampling

This is a better alternative, but not fool-proof. The sample is selected by taking every nth item on the list. This kind of sampling is widely used and can lead to good stratification, depending on how the list is ordered.

Systematic sampling from a list has less risk of bias when it is known that the list has no regularities. But one can rarely be quite sure of this. It is therefore safer to take a sample of size n by dividing the list into n equal parts (which can still provide good stratification) and randomly selecting one item from each part. Systematically picking every nth item merely avoids having to look up the random numbers.

Quota Sampling

Quota sampling is common in market or opinion surveys involving human populations. Here each interviewer is set certain quotas of different types of people to interview. like five white-collar workers aged 45–65, three non-working housewives aged under 35, etc. Then each interviewer personally selects the individuals to fill his quotas. This is where bias can occur despite additional rules and regulations about the selection. (E.g. the interviewers may be told to restrict their choices to specified districts, usually selected on a random or probability basis, and not to interview more than two people in any one street.) Quota sampling is therefore a form of stratified cluster sampling, but with the individuals in each stratum or 'cell' selected subjectively rather than at random.

Well-controlled quota sampling has given representative results for quite a wide range of topics, but it can only carry conviction in those limited situations where previous experience has shown it to work. For instance, with the purchases of some frequently-bought branded goods, the different brand-shares shown in the survey could be checked against known sales figures.

Quota sampling is often cheaper per interview than random sampling, the main alternative. But much of the high cost of random sample surveys comes from follow-up calls on informants not home initially. With quota sampling, such informants tend to be ignored. That is one reason it is cheaper. Random sampling would be less costly if no call-backs on initial 'not-at-homes' were made.

7.6 The Non-Sampling Approach

Formal statistical sampling—by random selection or otherwise—is not the most common way of collecting empirical data. For example, in studying the incidence of Symptom X among children under the age of 5 who have had German Measles, an investigator will seldom explicitly aim to take a sample that is representative of all such children in his home town or region, let alone in the country as a whole.

One reason is that the costs of building a comprehensive sampling frame are usually far too high. More important, one is seldom interested in just one overall result for the population as a whole—e.g. that 20 percent of all such children in the country had Symptom X. Even if such a result is broken down by region of the country and size of family, it still does not give enough information. One usually wants to compare results from very different kinds of circumstances, to see if generalizations and perhaps some revealing exceptions emerge.

Therefore a variety of different studies are needed. Instead of carrying out one big sample survey, relevant results are accumulated on a piecemeal basis, i.e. by studying children under 5 who have had German Measles under a variety of different conditions. Such studies will tend to be done by different people in different places, and at different points in time, usually with no master plan. Each will involve a small population, like all the records at the ABC hospital for the last year, or a sample from such a population. (The sample should be drawn along the lines already discussed in this chapter to avoid introducing unknown biases, but the specific population is usually of no great interest in its own right.)

Strong conclusions can still be drawn from a range of such apparently haphazard studies. If this year's results for the ABC hospital in the north-east and those for a private doctor's practice in the south-west 3 years ago both say that about 20 percent of the children had Symptom X, we have learned that neither region of the country, nor time, nor hospital versus private practice appear to affect the incidence of the symptom. If the same 20 percent result is obtained under a yet wider range of conditions (e.g. two studies in Ames, Iowa, and in Bangladesh), the generalization becomes still more powerful. It tells us more than any single statistical survey.

We would also learn something if different studies yielded *differing* results (e.g.

5 percent having Symptom X here, and 50 percent there). We would then know that there is no generalizable result and that Symptom X depends on a variety of other factors. We may even gain some clues of what the factors might be (e.g. if *hospital* results were consistently low). In such a case a statistical survey result of 20 percent for the country as a whole would be of limited relevance and potentially misleading, i.e. a 'bad average' of systematically different levels of incidence.

The process of generalization and inference used here is not statistical but empirical. It is discussed further in Part Six.

7.7   Discussion

A sample can never tell us more than if the whole population were measured. The aim in taking a sample is generally only to save time, effort, and money. But sampling introduces error. The advantages of using some form of random or probability sampling is that we can calculate how large the error is likely to be and make it small by taking a large enough sample. With non-probabilistic sampling, the amount of sampling error cannot usually be calculated.

The various elaborations of simple random sampling—such as multi-stage, stratified, cluster, and probability sampling—aim to reduce costs for a given level of accuracy. The technicalities of sampling are therefore of practical rather than fundamental concern. The detailed work tends to be done by people with specialized experience. But the rest of us need some appreciation of sampling procedures in order to understand the nature and limitations of the results.

CHAPTER 7   GLOSSARY

Area sampling

A form of cluster sampling where individuals are selected from specified areas (e.g. city blocks) without having a comprehensive sampling frame for the whole population.

Clustering

Selecting more than one item from a larger sampling unit, e.g. individuals from a selected district. (Sometimes the term is used only when *all* individuals are selected.) Saves money but usually entails somewhat larger sampling errors.

Convenience sampling

Selecting whichever items happen to be handy. Unlikely to give representative results.

Empirical generalization

Observed result which holds under a wide range of conditions.

Multi-stage sampling

Selecting a sample in stages; for example, first some towns, then some administrative districts in these towns, and then individuals in each district. Saves the costs of handling lists of all individuals in all towns.

Non-response	Refers to members of a sample who could not be measured.
Non-sampling errors	Errors in the observed data other than those due to sample selection (e.g. because of non-response).
Population	A specified set of people, objects, or measurements from which a sample is to be taken.
Posterior stratification	Weighting the readings in a sample to reflect the correct proportions of certain groups or strata in the population (e.g. different age-groups).
Prior stratification	Dividing the population into certain groups or strata *before* the sample is selected, in order to increase the sample's accuracy.
Probability sampling	Selecting items with unequal but known probabilities. The results can be made representative by weighting.
Quota sampling	Interviewers are told how many people of different types to interview, but select the individuals themselves.
Random sampling	Individual members of the population are selected with equal probabilities, but possibly subject to multi-stage selection, clustering, and/or stratification.
Random walk	A form of area sampling with individuals selected 'on the ground'. Usually not strictly random.
Randomized experiment	A study using random selection to form experimental and control groups.
Sample	A selection of objects or measurements from a specified population.
Sampling frame	A list of all members in the population to be sampled.
Sampling with replacement	Maintains a constant probability of selection by replacing chosen items in the population so they can be picked again.
Sampling without replacement	Selected items are *not* replaced in the population so that at each stage of selection the remaining items have a slightly higher probability of being selected. The effect on the sample results is usually very small.
Simple random sampling	The specific form of random sampling where there is no multi-stage selection, clustering, or stratification.
Stratified sampling	Using prior stratification to ensure that the correct sample sizes are selected from specified sub-groups of the population.
Systematic bias	Tendency for a result to differ consistently from the true value (e.g. generally too low).
Systematic sampling	Selecting every nth item on a list. Easy to do, but risks bias if the list has a pattern.
Variable sampling fraction	A form of probability sampling in which certain members of the population are selected in higher proportions (e.g. people with severe attacks of an illness). Needs weighting of results.
Weighting	Correction of a sample to improve its representation of the population sampled. Results for different members of the sample are multiplied by numerical 'weights' before the analysis starts.

CHAPTER 7   EXERCISES

7.1   The ages of twenty-five people working in a certain office are:

21   32   42   49   54
24   34   43   50   57
28   38   46   51   59
30   39   47   52   60
31   40   48   53   64

Select a sample of five people, and comment briefly, by a form of
(i) Systematic sampling.
(ii) Simple random sampling (use Tables 7.1 or A6).
(iii) Stratified random sampling.

7.2   You need to make certain chemical measurements of the sap of the trees in a large forest covering mountainous terrain. Briefly outline the problems of selecting a random sample of 200 trees.

7.3   In a factory's quality control inspection scheme, every hundredth item is checked. Last week the following results were obtained:

AM   5, 11, 7, 15, 5, 15, 10, 19, 9, 21, 15, 25, 14, 19, 16, 21.
PM   17, 17, 18, 23, 19, 22, 20, 24, 20, 23, 20, 25, 18, 19.

It is proposed to reduce the sample size by inspecting every two-hundredth item. Comment on how this might be done.

7.4   What is the difference between stratified random sampling and quota sampling?

7.5   It was said in the main text that people living in the same district, street, or household are often relatively similar, thus reducing the value of selecting more than one individual from such a cluster. Discuss whether this is likely to be so, given for example that a household will usually include people of both sexes and different ages?

7.6   'A pack of cards is shuffled and a card is drawn at random.' Can this be true?

CHAPTER 8

How Sample Means Vary

A random sample will usually not be exactly representative of the population from which it was selected. For example, the sample mean m will probably differ from the population mean μ (Greek m or 'mu'). The difference will often be small, but we cannot measure it directly since we do not know μ. If we did, we would not have needed the sample.

An indirect way of assessing the difference comes from seeing how different the means of *other* such samples would be. The less these vary, the more likely it is that our sample mean gave an accurate answer.

We therefore have to consider the distribution of the means of all the possible random samples of size n from the same population. This is called the *sampling distribution of the mean*. To describe this distribution we need to know its shape, its mean, and its scatter. Sections 8.2–8.4 show how we can estimate these three features from the information contained in the one sample actually taken.

Small samples have more complicated sampling distributions, as noted in Section 8.5. Determining the sample size required for a given level of accuracy is discussed in Section 8.6.

Any other statistical summary measure, like the variance or standard deviation, will also vary from sample to sample and hence have a sampling distribution. This is discussed in Section 8.7.

8.1 A Series of Samples

Suppose we want to know how many purchases of Corn Flakes were made by households in the town of Southampton in the previous 6 months. Instead of measuring the purchases of all households in Southampton, we select a random sample and ascertain their half-yearly purchases of Corn Flakes by some suitable method. To keep the illustration simple, we consider a small sample of ten households. The number of Corn Flakes purchases they made are as follows:

First sample: 3, 0, 0, 4, 2, 0, 0, 2, 0, 12   Average: 2.3

Now suppose we take a second sample of ten households. This gives different results:

Second sample   4, 1, 1, 0, 17, 1, 0, 3, 1, 2   Average: 3.1

The mean is 3.1 instead of 2.3.

Other samples would give still different results and different sample means. Table 8.1 shows the means of twenty such samples, arranged in order of size.

TABLE 8.1   The Means of Twenty Random Samples of Ten Households
(Average purchases of Corn Flakes, ordered by size)

Means of Samples of 10
.6, 1.7,   2.0, 2.0, 2.0, 2.2, 2.3, 2.3,   3.0, 3.1, 3.2, 3.4, 3.6,   4.1, 4.3, 4.6,   5.0, 5.2, 5.2,   6.0

The twenty sample means range from .6 to 6.0, but most of them are about 2 or 3 and tell much the same story. This is so even with samples as small as ten.

If we took all possible samples of ten households from Southampton, we would have the full distribution of sample means for random samples of $n = 10$. Table 8.1 already gives us some indication of what this distribution would look like. We now examine its properties more fully: its likely shape, its mean, and its scatter.

8.2   The Shape of the Distribution of Sample Means

Usually the sampling distribution of the mean is approximately Normal. This is always so when the variable measured has a roughly Normal distribution in the population itself (as we can judge roughly from our sample, and from any previous experience of similar data). But even when the measured variable is highly skew in the population, the sample means will have a nearly Normal distribution as long as the sample size n is more than about 100.

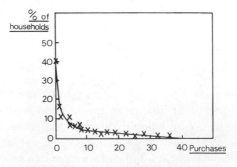

**FIGURE 8.1   Half-Yearly Corn Flakes
Purchases by 200 Households**
(with free-hand curve)

The Corn Flakes data illustrate this. Here the distribution of the individual readings in the population is highly skew. Figure 8.1 shows the purchases for
individual Southampton households. The distribution is reverse-*J*-shaped: 40
percent of the households were non-buyers, quite a few were light buyers, and
hardly any were heavy buyers.

But the distribution of the means of samples of $n = 10$ households is far
less skew. Figure 8.2A shows it has a somewhat long 'tail' to the right, but a hump
almost in the middle. If we took samples of $n = 100$ from that population, the
distribution of their means would get closer to a Normal shape, as Figure 8.2B
shows.

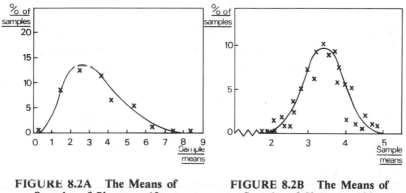

FIGURE 8.2A   The Means of   **FIGURE 8.2B   The Means of**
Samples of Size $n = 10$ **Samples of Size $n = 100$**

(Each figure is based on 200 samples, with free-hand curve)

The tendency for the sampling distribution of the mean to be approximately
Normal has a strong theoretical basis in the Central Limit Theorem (Chapter 6).
This says that a Normal distribution must result for repetitive measurements if
each is made up of a large number of small elements which are independent of
each other and additive. The sampling distribution of the mean meets these
requirements:

(i) Taking the means of all possible samples of size n is a form of repetitive
measurement.

(ii) The measurements can be made large in number by taking a large enough
sample size n.

(iii) The samples are composed of individual readings, i.e. small elements.

(iv) In random sampling the measurements are selected independently of each
other.

(v) They are added together when calculating the mean $m = \text{Sum}(x)/n$.

8.3   The Mean of the Sampling Distribution

We now consider the *mean* of the sampling distribution. This equals the mean μ of the whole population that is being sampled—a very simple result.

For our numerical example, the average number of half-yearly Corn Flakes purchases in Southampton as a whole was $\mu = 3.4$ per household (although in practice one would not know this). Successive random samples of ten households have different means, as we saw in Table 8.1. If we took all possible random samples of ten, the average of all their means m would work out at 3.4—equal to the population mean μ. In other words, the sample mean m will on *average* give the right answer μ.

To illustrate how this works, consider a small skewed population of just three measurements:

$$2, 3, 7.$$

The mean of this population is $\mu = (2 + 3 + 7)/3 = 4$. There are nine possible samples of size $n = 2$ when sampling 'with replacement' (see Section 7.4):

2 & 2,   2 & 3,   2 & 7,   3 & 2,   3 & 3,   3 & 7,   7 & 2,   7 & 3,   7 & 7.

The means of these samples are

$$2, \quad 2\tfrac{1}{2}, \quad 4\tfrac{1}{2}, \quad 2\tfrac{1}{2}, \quad 3, \quad 5, \quad 4\tfrac{1}{2}, \quad 5, \quad 7,$$

or ordered by size

$$2, \, 2\tfrac{1}{2}, \, 2\tfrac{1}{2}, \, 3, \, 4\tfrac{1}{2}, \, 4\tfrac{1}{2}, \, 5, \, 5, \, 7.$$

This constitutes the sampling distribution of the mean, since with random sampling all nine samples must come up equally often in the long run.

The mean of this distribution is

$$(2 + 2\tfrac{1}{2} + 2\tfrac{1}{2} + 3 + 4\tfrac{1}{2} + 4\tfrac{1}{2} + 5 + 5 + 7)/9 = 4$$

which equals the mean μ of the population.

The same argument applies for larger populations and for larger samples (and also for sampling without replacement). We therefore now know that the overall mean of the distribution of sample means will be equal to the population mean μ and that the shape of the distribution will usually be approximately Normal.

8.4   The Standard Error of the Mean

Finally we need to consider the scatter of the distribution of sample means. We want to know how far from the population mean μ any particular sample mean is likely to be. This scatter is usually expressed in terms of the standard deviation.

The standard deviation of the distribution of the sample means is usually given a special name: *the standard error of the mean* (or standard error for short). The name implies that it is the average or usual error of the observed mean m due to its being based only on a sample.

To calculate the numerical value of the standard error we could in principle take all the possible samples of size *n* from the given population, compute their means, and calculate the standard deviation of these different means. That is how the standard error is defined. (We looked at such distributions of sample means in Section 8.1.)

But in practice we only have a single sample and cannot do these calculations. There is however a theoretical formula which allows us to estimate the value of the standard error from the single sample we have.

A Theoretical Formula

The theoretical formula for the standard error of the mean is expressed in terms of the scatter of the individual readings in the population. Their standard deviation is denoted by σ (the Greek s or 'sigma'). For simple random samples of size *n*, the formula is

$$\text{Standard error of the mean} = \frac{\sigma}{\sqrt{n}}.$$

A feature of this formula is that it uses the (unknown) standard deviation of the *individual readings* in the population, and not the (unknown) means of all possible samples of size *n*. The mathematical derivation of this formula is outlined in more advanced texts, but we can readily see that it makes sense in two respects.

Firstly, the formula says that the smaller σ, the scatter of the individual readings, the smaller the standard error will be. Thus if the individual readings in the population do not differ much, the means of different samples also will not differ much.

Secondly, the formula says that the larger the sample size *n*, the smaller the standard error will be. This makes sense because the more measurements we take, the closer we get to the true population value. Figure 8.3 illustrates this effect for

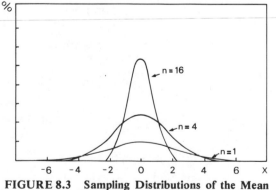

FIGURE 8.3   Sampling Distributions of the Mean for Samples of *n* = 1, *n* = 4 and *n* = 16 from a Normal Distribution with Zero Mean

sampling distributions of the means for $n = 4$ and $n = 16$ readings from a Normal population with $\mu = 0$ and $\sigma = 3$. It also compares them with the distribution of single readings, as if one took samples of $n = 1$.

However, the reduction in the standard error is not directly proportional to the sample size. The standard error does not vary with $1/n$, but with $1/\sqrt{n}$. Thus if the sample size is multiplied by 4 (e.g. from $n = 4$ to $n = 16$), the standard error is only halved (i.e. it is multiplied by $1/\sqrt{4} = \frac{1}{2}$). The reason is that each additional reading in the larger sample tells us less than the preceding ones. Even from the initial small sample we already have a fair idea where the population mean is.

Estimating the Standard Error

In practice the standard error formula $\sigma/\sqrt{n}$ is still of no direct use since we do not know the value of the population standard deviation σ. We have not measured the whole population; we have only measured a single sample of n readings.

To get around this problem, we *estimate* the value of σ from the standard deviation of the readings in the sample, denoted by s. The formula for the *estimated* standard error of the mean is therefore

$$\text{Estimated standard error of the mean} = \frac{s}{\sqrt{n}}.$$

This is the formula one uses in practice. The fact that we have substituted the sample value s for σ makes almost no difference to any conclusions, unless the sample size n is less than about 10. (Such cases will be discussed in Section 8.5.)

To illustrate the numerical calculation of $s/\sqrt{n}$, we return to the Corn Flakes purchases made by the first sample of ten Southampton households. The readings were 3, 0, 0, 4, 2, 0, 0, 2, 0, 12, and their mean is $m = 2.3$. So their standard deviation is

$$s = \sqrt{\left(\frac{\text{Sum } (x - 2.3)^2}{(n - 1)} \right)} = 3.73.$$

The estimated standard error of the mean is therefore

$$\frac{s}{\sqrt{n}} = \frac{3.73}{\sqrt{10}} = 1.18.$$

The full formula for the estimated standard error of the mean based on a sample of n is

$$\frac{s}{\sqrt{n}} = \frac{\sqrt{\left(\dfrac{\text{Sum } (x - m)^2}{(n - 1)} \right)}}{\sqrt{n}}.$$

The Interpretation of the Standard Error

The standard error of the mean $(\sigma/\sqrt{n})$ summarizes how far the means of all possible samples of size n would lie from the population mean μ. As noted earlier, the shape of the distribution of sample means is approximately Normal (as long as the variable being measured in the population is roughly Normal *or* the sample size is one hundred or more). Therefore we can describe the distribution of the sample mean as follows:

68 percent of all possible sample means will lie within $\pm1\sigma/\sqrt{n}$,
95 percent of all possible sample means will lie within $\pm2\sigma/\sqrt{n}$.

When we use the *estimated standard error* this description still applies, but only if

(i) n is greater than about 10 when sampling from roughly Normal populations *or*

(ii) n is greater than about 100 when sampling from highly skew populations.

If these conditions are satisfied, substituting $s/\sqrt{n}$ for the unknown $\sigma/\sqrt{n}$ hardly matters: it is still approximately true that about 68 percent of all possible sample means will lie between $\pm1s/\sqrt{n}$ from μ.

However, the Normal approximation does *not* hold if the sample size is less than ten with Normal populations. Then the procedures outlined in the next section have to be used.

8.5   Student's *t*-Distribution

We saw above how the Normal distribution can be used to give a good approximate description of the distribution of the sample means. A more precise description is given by Student's *t*-distribution. (The mathematics of the *t*-distribution were first derived by W. S. Gossett, who published under the pseudonym 'Student'.)

To use the *t*-distribution we first have to consider what is called the *t*-ratio or *t*-variable:

$$t = \frac{m - \mu}{(s/\sqrt{n})}.$$

As the equation shows, t is defined as the difference between the observed sample mean m and the population mean μ, divided by the estimated standard error. (This looks like a standardized variable as discussed in Chapter 2, but the divisor here is the *estimated* standard deviation, not the true one.)

The reason we calculate the *t*-variable is that its theoretical distribution happens to be known. This is Student's *t*-distribution. It gives the proportion of times different values of the *t*-ratio are likely to occur for all possible samples of a given size from a Normal or nearly-Normal population.

The *t*-distribution can take different shapes, depending on the sample size n. Its precise form depends on the number $(n-1)$, which statisticians term the 'degrees of freedom'. Table 8.2 sets out some values of the *t*-distribution for different degrees of freedom.

TABLE 8.2   Values of the *t*-distribution

(Limits on either side of the mean μ which contain 90%, 95%, 99% and 99.9% of the values of the *t*-variable for various degrees of freedom)

	Degrees of freedom									
	1	2	3	4	5	10	15	20	30	Large*
90%	6	3	2.4	2.1	2.0	1.8	1.8	1.7	1.7	1.6
95%	13	4	3.2	2.8	2.6	2.2	2.1	2.1	2.0	2.0
99%	64	10	5.8	4.6	4.0	3.2	2.9	2.8	2.7	2.6
99.9%	640	32	12.9	8.6	6.9	4.6	4.1	3.8	3.6	3.3

* As for the Normal Distribution

Suppose we had a sample of $n = 11$ readings from a given population. Then we would look in the column for $(n - 1) = 10$ degrees of freedom. The first figure of 1.8 would tell us that for 90 percent of all possible random samples of size 11 from that population, the values of the *t*-ratio would lie between ± 1.8. Similarly, 95 percent of the values of the *t*-ratios would lie between ± 2.2.

If we transpose the terms of the *t*-ratio, the last statement says that 95 percent of all the possible sample means m would lie between $\pm 2.2\ (s/\sqrt{n})$ on either side of the population mean μ. Thus we have two equivalent statements:

$$95\% \text{ of all values of } t = \frac{m - \mu}{s/\sqrt{n}} \text{ lie between } \pm 2.2,$$

or

$$95\% \text{ of all values of } m \text{ lie between } \mu \pm 2.2\ (s/\sqrt{n}).$$

The results also summarize the tails of the sampling distribution of the mean. For example, 5 percent of all values of m for samples of $n = 11$ would lie *outside* the limits $\mu \pm 2.2\ (s/\sqrt{n})$.

The Normal Approximation

If we look at the five right-hand columns of Table 8.2 we see that the values in each row are mostly pretty similar. They imply for example that 90 percent of sample means would lie within about ± 1.7 standard errors on either side of μ, and that 95 percent would lie within about ± 2.0 to 2.2 standard errors of μ.

These figures are close to those of a Normal distribution. In fact, the figures given in the 'Large' column (for n well over 30) are identical to those for the Normal distribution. But even for samples with only eleven readings and ten degrees of freedom the *t*-distribution takes much the same form as a standardized Normal distribution (except in the extreme tails, say for .01% of samples).

This makes description of the distribution of sample means much easier. As noted earlier, for sample sizes greater than about ten we can use the more familiar scatter limits of the Normal distribution.

The reason this approximation holds is that for sample sizes greater than 10 from a nearly Normal population, substituting the estimated standard error $s/\sqrt{n}$ for the unknown $\sigma/\sqrt{n}$ has hardly any effect on the nature of the sampling distribution. It still follows a roughly Normal form. But when the sample size is ten or less, the Normal approximation does *not* hold—as Table 8.2 shows. In those cases one must use the *t*-distribution to describe the sampling distribution of the mean. (This requirement for n to be greater than 10 should not be confused with the requirement that when sampling from a highly *skew population* the sample size has to be greater than about $n = 100$ to get an approximately Normal distribution of the sample means.)

Student's *t*-distribution was a pioneering effort in the history of statistics because for the first time problems of small samples were dealt with accurately. But the practical implication is that we can ignore the problem, except for *very* small samples. And in practice such samples are rare.

Degrees of Freedom

The term 'degrees of freedom' which arises with the *t*-distribution also comes up in other statistical analyses (like the 'analysis of variance' referred to in Chapter 4). The basic concept is that n variable quantities can vary in n different ways. One can thus say that the data have n 'degrees of freedom'. However, each time we calculate a parameter from the data, we use up one of these degrees of freedom.

To illustrate, suppose we have three readings of x, taking the values:

$$x: \quad 1, \ 3, \ 5$$

The mean m is 3. The deviations from the mean $(x - m)$ are

$$(x - m): \quad -2, \ 0, \ 2.$$

As always, the deviations from the mean must add to zero, i.e. Sum $(x - m) = 0$.

Thus the values of the first two deviations (-2 and 0) dictate that the third value *must* be 2. Only two of the deviations can vary independently. Calculating the mean has taken away one degree of freedom from the data.

This generalizes for any number of readings n. Having calculated the mean, we have effectively used up the independence of one of the readings. Since there is no reason to identify one specific reading as being 'used up', we say that one 'degree of freedom' has been used in the data set.

In general, one degree of freedom has to be subtracted for each parameter calculated for the data. Thus when it comes to computing the variance or standard deviation of a data set, one degree of freedom has already been used to calculate the mean, so there are only $(n - 1)$ degrees of freedom left. That is why the figure $(n-1)$ is used as the divisor in those formulae. It leads to many mathematical simplifications in the statistical theory.

8.6   Determining the Sample Size

The theoretical standard error formula $\sigma/\sqrt{n}$ can help to determine the sample size needed to achieve a given level of accuracy in a sample survey or experimental study.

Suppose that the standard deviation of the population to be sampled is about five, i.e. $\sigma \doteqdot 5$. If our aim is that the standard error of the sample mean m should be .5, a sample of size $n = 100$ would be needed to ensure that $\sigma/\sqrt{n} = .5$. If accuracy were required to within a standard error of .1, a sample size of about $n = 2500$ would be needed, since $5/\sqrt{2500} = .1$. Thus a five-fold reduction of sampling errors would require a twenty-five-fold increase in sample size.

For the above calculations we would in principle need to know the value of the population standard deviation σ even before the sample data were collected. Strictly speaking this is impossible. However, there is often some previous work in the same area or a pilot-study to provide a rough idea of the likely value of the scatter of the readings in the population. One might know that σ will probably be between 4 and 6 rather than between 10 and 20. That is enough information to calculate roughly what sample size is needed, i.e. somewhere between 1600 and 3600 to get a standard error of .1. That may seem a wide range, but the important thing is to know that a sample size of roughly 2000 to 3000 would be needed, instead of 200 or 20 000.

In practice, there is usually no clear-cut answer to how large a sample is needed. Most studies involve a number of different variables which tend to have different degrees of scatter in the population. Most studies also have more than one purpose and differing levels of precision may be required for each. Nonetheless, the standard error formula $\sigma/\sqrt{n}$ helps in planning sample investigations.

8.7   Other Sampling Distributions

Any statistical summary measure calculated for a sample will vary from sample to sample and hence have a sampling distribution. For instance, the standard deviation of the sample readings will have a distribution of values over all the possible random samples of a given size n.

Most sampling distributions tend to have an approximately Normal shape if the sample size n is large enough. For the sampling distribution of the mean, 'large' usually means $n = 100$, even for very skew populations. But for many other summary measures the sampling distributions are approximately Normal only if the samples are as large as $n = 1000$ or more.

Since in practice samples are often not that large, it follows that there are many sampling distributions which are *not* Normal. One can then no longer use the simple interpretation of a standard error, i.e. that 95 percent of the values from different samples lie between ±2 standard errors of the population value. Instead one has to refer to tables which set out the sampling distribution in question (like Table 8.2 for the *t*-distribution). Some of the more important cases of this are described in more advanced statistical texts. We shall come across examples of the F- and χ^2-distributions in Chapter 10. But there are also cases where no theoretical solution exists.

Sampling Theory for Other Types of Sampling

Until now we have considered only sample data collected by simple random sampling, where every item in the population is given the same probability of being selected, with no stratification, multi-stage sampling, or clustering. But most sampling is not done this way. In sample surveys, for example, some form of stratified multi-stage cluster sampling is more typical, as noted in Chapter 7. This affects the sampling errors that occur, and hence the resulting sampling distributions. For example, multi-stage cluster sampling will usually increase the size of sampling errors compared with those of a simple random sample of the same size. Conversely, stratification may decrease the size of sampling errors.

Unfortunately, it is often hard or even impossible to establish precisely how much these methods will affect the sampling errors. Standard error formulae appropriate for simple random sampling are therefore widely used, or misused, as if they applied to these more complex forms of sampling.

Sometimes allowance is made for a so-called 'Design Factor'. This is an estimate of how much larger the variance of the sampling errors will be than for simple random sampling. Such estimates may be based on comparable previous empirical experience, on theory, or on guess-work.

8.8   Discussion

A sampling distribution describes how a particular summary measure of a sample varies for all the possible samples of size *n* from the given population. For simple random samples, the sampling distribution of the *mean* has three major properties:
- It tends to be Normal, except for very small samples or for samples less than $n = 100$ from highly skew distributions.
- Its mean is equal to the population mean μ.
- Its standard deviation is numerically equal to $\sigma/\sqrt{n}$ and can be estimated by the standard error formula $s/\sqrt{n}$ (where *s* is the observed standard deviation of the *n* individual readings in the sample).

In practice we only observe a single sample from a given population. But the formula $s/\sqrt{n}$ for the estimated standard error of the mean lets us establish from that single sample how the means of all other possible samples would differ from the population mean μ. This is a remarkable achievement. It is possible in random sampling because each item in the sample is selected independently of all the other items. Hence we can use probability mathematics to calculate how much different samples would vary.

A marked simplification also arises from the fact that we can use the observed value of the sample standard deviation (instead of the theoretically required but unknown population value σ) unless the sample size *n* is very small. In the latter case the *t*-distribution provides the solution, at least when sampling from a population where the distribution of the measured variable follows a nearly Normal distribution.

Knowing the standard deviation of the sampling distribution of the mean would be of limited use if we did not also know the shape of the distribution. Here the Central Limit Theorem brings about yet another major simplification for random sampling: in general, the sampling distribution of the mean is roughly Normal. This tells us how we can interpret the standard error: i.e. about 95 percent of all possible sample means will lie within ±2 standard errors of the population mean μ. This is a very simple result. It is highlighted by the fact that the sampling distributions for summary measures other than the mean are generally much more complex.

The special contribution of the statistical sampling theory we have discussed in this chapter is that we can use a single observed sample to assess how all the different samples of the same size would behave. We shall see in Chapters 9 and 10 how this knowledge is used by statisticians to make more specific inferences from the observed sample data to the population in question.

CHAPTER 8   GLOSSARY

Degrees of freedom	The sample size *n* minus the number of parameters already calculated from the sample.
Design factor	Estimates the increase in the estimated standard error for random sampling due to having used a selection procedure other than *simple* random sampling.
Estimated standard error	$s/\sqrt{n}$, the value of the standard error as estimated from the standard deviation of the observed sample.
m	The sample mean.
μ	The population mean (Greek m or 'mu').
Sampling distribution of the mean	The distribution of the means of all possible random samples of size *n* from a given population.
Standard error	Shortened name for the standard error of the mean.
Standard error of the mean	A special name for the standard deviation of the sampling distribution of the mean. Represents the average size of the error between *m* and μ due to *m* being based on a sample.
Standardized variable	The observed readings minus their mean and divided by their standard deviation. A standardized variable therefore has a mean of zero and a standard deviation of 1.

s	Standard deviation of the individual readings in an observed sample.
σ	The standard deviation of the individual readings in a population (Greek s or 'sigma').
$s/\sqrt{n}$	The estimated standard error of the mean for an observed sample of n readings.
$\sigma/\sqrt{n}$	Formula for the true but unknown standard error of the mean.
Student's *t*-distribution	The distribution of the sample means m when expressed in the form of the standardized t-ratio. It is useful for small sample sizes.
t-ratio or *t*-variable	$(m - \mu)/(\sqrt{s}/n)$, i.e. the difference between the sample and population means, divided by the estimated standard error of the mean.

CHAPTER 8   EXERCISES

8.1   A certain small population consists of the following four readings:

$$0, 2, 4, 6.$$

(a) Using sampling with replacement, find all sixteen possible samples of two readings and work out the sample means. Set out the results in summary form. What would the resulting distribution be called? Comment on its shape.

(b) Work out the standard deviation of the sixteen sample means. What is this usually called? Compare it with the standard deviation of the four readings in the population and comment. (Use the divisor n throughout these calculations, because we are sampling with replacement.)

8.2   A random sample of $n = 20$ has given the following observations:

$$2, 3, 3, 4, 4, 5, 5, 5, 6, 6,$$
$$6, 6, 7, 7, 7, 8, 8, 9, 9, 10.$$

(a) Work out the mean m, standard deviation s, and the estimated standard error of the mean.

(b) What does the estimated standard error tell you?

8.3   (a) The means of all possible samples of size n from a given population have a standard deviation equal to $\sigma/\sqrt{n}$. How can we use this theoretical result to describe the scatter of the means of the different samples if we do not know the value of σ, the standard deviation in the population?

(b) Why do we have to use the t-ratio $(m - \mu)/(s/\sqrt{n})$ for small samples?

(c) For samples of $n = 6$, what percentage of sample means m lie more than ± 4 times the standard error from the mean?

8.4   (a) The standard deviation of a certain population is about 5 units. For simple random sampling, how large a sample is needed if the standard error of the mean is to be 1? How large for a standard error of 0.1?

(b) How can one know the standard deviation of the population before having taken a sample from it?

8.5 (a) In a sample of size $n = 900$, the observed standard deviation of the individual readings is 3.0. What is the estimated standard error of the mean if simple random sampling was used?

(b) The data were in fact collected by a form of stratified multi-stage sampling which has a 'Design Factor' of 1.9. What is the estimated standard error of the mean?

CHAPTER 9

Estimation

When faced with a sample, our interest centers on the population from which it was selected. What does the sample tell us about that population?

Arguing from the sample to the population is called *statistical inference*. In this chapter we discuss two aspects of this task:
  (i) Estimating the numerical characteristics of the population from those of the sample; and
  (ii) Assessing the accuracy of these estimates, usually in terms of 'confidence limits'.

We shall generally assume that the sample was selected from a specified popula tion by simple random sampling. The general principles are the same for other forms of probability sampling, but the calculations are more complex.

9.1 Estimation

The first task is to estimate the characteristics of the population. For example, given a sample of n readings with a mean $m = 9$ and a standard deviation $s = 4$, what can we say about the corresponding population values or parameters, μ and σ? (A numerical population value like a mean or standard deviation is usually called a *parameter*, while the corresponding value for the observed sample is called a *statistic*.)

Common sense suggests that we can simply use the sample values, m and s. For the *mean*, this is so. The sample mean $m = 9$ is the best estimate of the population mean μ, by any criterion.

But for other parameters, the observed sample value or statistic may not be the right or best estimate. A dramatic illustration is given by the range. The range of readings in a sample is nearly always much smaller than the range of the population.

One possible criterion in judging a good estimator of the population value is lack of statistical bias—i.e. giving the right answer on average across all possible samples. This works for the mean m but not for the variance s^2. If the latter is defined as the straight average of the n squared deviations (i.e. as

Sum $(x - m)^2/n$), the average value of this across all possible samples will be slightly smaller than the population variance σ^2. This systematic error or 'bias' in the sample estimate is eliminated by using the divisor $(n - 1)$ instead of n. (This is one of the reasons for using the 'degrees of freedom' as a divisor, as was mentioned in Chapter 2.)

But if we accordingly define the sample standard deviation as $s = \sqrt{\{\text{Sum} (x - m)^2/(n - 1)\}}$, this is not an unbiased estimator of the population standard deviation σ. (The average of the square roots of a set of numbers is not equal to the square root of the average.) Yet in practice we generally use this definition of s, partly because the bias is very small and partly because there is no unambiguously better alternative.

Another estimating criterion, often used in more complex situations, is called the 'maximum likelihood' principle. This chooses as an estimate that value which, if true, would give the highest probability (or 'maximum likelihood') that the sample in question would actually have been observed.

For the mean, the maximum likelihood principle leads again to the sample mean m as the 'best' estimate of the population mean μ. For example, a sample with a mean of 9 is more likely to have been selected from a population with a mean of 9 than from one with a mean of 8. Hence 9 is the 'most likely' population value.

For the variance and standard deviation, the mathematics of the maximum likelihood principle lead to the divisor n rather than $(n - 1)$. The resulting estimators are then not unbiased—they no longer give the population value on average. But lack of bias is not an absolute requirement.

This illustrates the complications which can arise in estimating a population parameter from sample data. Nonetheless, estimation is straightforward for most of the basic statistical measures used in this book, like the mean, median, mode, and standard deviation (or the correlation and regression coefficients of Part Four). Following common sense, the value obtained for the random sample is taken to be the estimate of the value in the population, perhaps with minor modifications like the divisor $(n - 1)$ rather than n.

9.2 Confidence Limits

Having chosen a sample estimate of a particular population value, we now have to assess how accurate this estimate is. Because we are dealing with random samples, the answer will be in the form of probabilities, telling us how *likely* we are to be wrong and by how much.

For example, suppose we want to estimate the population mean μ from a random sample of $n = 100$ readings with mean $m = 9$, standard deviation $s = 4$, and hence an estimated standard error of $s/\sqrt{n} = .4$. Our *estimate* of μ is the sample mean 9. But almost certainly the population value is not 9. How far out is our estimate likely to be?

Roughly the answer is that m will hardly ever be wrong by more than 2 or 3 times its standard error of .4, i.e. by more than .8 or 1.2 in our example, which is about ± 1.

A more precisely-worded answer can be obtained by calculating *confidence limits* for the estimate. The argument is fairly complex. We start with the knowledge that with n as large as one hundred, the distribution of sample means will be virtually Normal, with an unknown mean μ and an estimated standard deviation $s/\sqrt{n} = .4$. It follows that 95 percent of all possible samples will have means m which will lie between $(\mu - 2s/\sqrt{n})$ and $(\mu + 2s/\sqrt{n})$, i.e. between $(\mu - .8)$ and $(\mu + .8)$ in our case. If our observed sample were one of these samples, its mean $m = 9.0$ must differ from μ by less than .8. Hence the unknown μ must differ from $m = 9$ by less than .8, i.e. it must lie between 8.2 and 9.8.

This kind of statement will however only be true for 95 percent of all the possible samples, i.e. with a probability of .95. This probability is commonly referred to as one's 'confidence' of being right in saying that the two-standard-error limits on either side of the observed sample mean m would contain the unknown population mean μ. These limits (here 8.2 and 9.8) are referred to as the '95 percent confidence limits' for the given sample.

The result is more complex than it may appear on the surface. It is not a probability statement about the unknown but fixed parameter μ, but about the results of different random samples. For different samples of size n from the same population, the two-standard errors limits would not be the same. In our example, another sample of $n = 100$ from the same population might have a mean of 9.4 and a standard error of .3, leading us to calculate 95 percent confidence limits of 8.8 and 10.0 instead of 8.2 and 9.8. We therefore have to make the convoluted assertion that if for any one sample we were to say that the population mean lies between its own particular 95 percent confidence limits, then for 95 percent of all samples the *corresponding* type of statement would be correct (i.e. between 8.2 and 9.8 for the first sample, between 8.8 and 10.0 for the second, between 8.1 and 9.7 say for the third, and so on).

This may seem almost intolerably complex. But in practice the confidence limits for most samples will be numerically similar, since the means of most random samples from a given population are fairly similar! It follows that the rough-and-ready interpretation of the confidence limits for our first (and only) sample—that the population mean μ lies in the range of 8.2 and 9.8 with a probability of .95—will be close to the truth. The choice is between making a statement which is true but so complex that it is almost unactionable, and making one which is much simpler but not quite correct. Fortunately, the effective content of the two kinds of statement is generally similar.

The Level of Confidence

Confidence limits with other probability values can be derived by the same kind of argument. If the sampling distribution is approximately Normal with a certain estimated standard deviation, the probability levels in Table 9.1 can be used. For

**TABLE 9.1   Descriptive
Characteristics of a Normal Sampling Distribution**
$(s/\sqrt{n}$ is the estimated standard
error)

Distance from the mean	% of readings within the stated limits
$\pm$ .6 s/$\sqrt{n}$	45%
$\pm$ 1.0 s/$\sqrt{n}$	68%
$\pm$ 1.6 s/$\sqrt{n}$	90%
$\pm$ 2.0 s/$\sqrt{n}$	95%
$\pm$ 2.6 s/$\sqrt{n}$	99%
$\pm$ 3.0 s/$\sqrt{n}$	99.7%
$\pm$ 3.3 s/$\sqrt{n}$	99.9%
$\pm$ 3.9 s/$\sqrt{n}$	99.99%

example, we can say that the population mean μ will lie in an interval between
± 3.3 times the estimated standard error $s/\sqrt{n}$ on either side of the observed
sample mean m and expect to be right in 99.9 percent of all possible samples. The
risk of being wrong is only 1 in a 1000.

For our numerical example with an observed sample mean of 9 and an
estimated standard error of .4, our expectations for the population mean μ are
therefore that it would lie between

8.2 and   9.8 with 95 percent confidence,
8.0 and 10.0 with 99 percent confidence,
7.7 and 10.3 with 99.9 percent confidence.

The risk of being wrong decreases sharply (from 5 percent to .1 percent, a
factor of 50). But the width of the confidence limits increases relatively little, from
1.6 to 2.6 (that is 8.2–9.8 versus 7.7–10.3). Thus for the 5 percent of all samples
where μ lies outside the two-standard-error confidence limits, most of the time it
will lie only *just* outside these limits. In our example two small increments of 0.5
units each (i.e. from 7.7 to 8.2, and from 9.8 to 10.3) will cover all but 1 in 1000
samples. Even in these rare cases, μ will mostly be *just* beyond these wider limits.
We can therefore be pretty sure that the population mean lies roughly between 8
and 10, or fractionally outside these limits.

9.3   Small Samples

We saw in Chapter 8 that when the sample size is 10 or less and one is using the
estimated standard error $s/\sqrt{n}$, the sampling distribution of the mean cannot be

assumed to be Normal. To set confidence limits we then need to work with the *t*-distributions of Table 8.2, referring to the standardized *t*-ratio

$$t = \frac{m - \mu}{s/\sqrt{n}}.$$

To illustrate, for a sample of $n = 6$ and thus $(n - 1) = 5$ degrees of freedom, the value of *t* for a 95 percent probability is 2.6. This means that the *t*-ratio for 95 percent of all possible samples of size 6 will lie between -2.6 and $+2.6$.

In these cases $(m - \mu)$ will therefore lie between $\pm 2.6s/\sqrt{n}$. Hence the 95 percent confidence limits for μ are from $(m - 2.6s/\sqrt{n})$ to $(m + 2.6s/\sqrt{n})$.

The confidence limits are wider than they would be for a large sample with the same estimated standard error. This is because the small sample size makes the estimate of the standard error rather uncertain.

9.4   Prior Knowledge

In recent years the so-called Bayesian approach has been developed to try to improve inferences from sample data by explicitly taking into account any prior knowledge one may have of the situation. For example there may be results from previous studies. Faced with a sample mean $m = 9$, our estimate of the population mean could be influenced by knowing that in a previous study the mean had been 6. Or one may have some hunch based on experience.

The basic step in the Bayesian approach is to try to translate one's previous information or hunch into probabilities about the likely value of the unknown population mean μ. For example, before the new data are collected one would have to attach probabilities to the possible values of μ, e.g. that there is a zero probability that the population mean will be about 3 or less; a .1 probability that μ will be about 4; a probability of .2 that μ will be about 5; a peak probability of .5 that μ will be about 6 (as in the earlier study); and so on. These are the 'prior probabilities' of μ.

The 'posterior probabilities' of the likely values of μ are then obtained by combining these prior probabilities with the information contained in the newly observed sample. The technique is to use Bayes' Theorem, which was referred to in Section 6.6. This says that given a sample mean $m = 9$, the probability of the population mean μ taking any particular value, say 5, is proportional to the prior probability that μ would be 5, multiplied by a factor reflecting the probability that a sample result of $m = 9$ would have been obtained if the population mean had been 5. Instead of simply accepting the sample result of 9 as the best estimate of the unknown μ, it is adjusted downward if one believed beforehand that lower values were more likely.

The idea of adjusting the conclusions to be drawn from the new sample in the light of one's prior probabilities makes sense at first. But the approach is not very widely used. One reason is that unless the new sample is very small (which it seldom is), the sample evidence will be overwhelming and the procedure becomes trivial.

Another difficulty is fixing on the prior probabilities. An attraction of the Bayesian approach is to allow one to think in terms of the probability of the unknown population mean μ taking different values. This is like saying in everyday life that 'The probability of rain tomorrow is .6.' As noted in Chapter 6, such prior probabilities are generally referred to as 'subjective probabilities' and are supposed to reflect one's personal beliefs. But few people (including scientists and statisticians) think about things in numerically precise probabilistic terms. Adjusting the observed sample information in the light of a prior probability which one has not really been thinking about is, therefore, not such an obvious thing to do.

Nonetheless, we do often have prior knowledge. It is usually difficult to express it in terms of probabilities, but it can help us to choose hypotheses to test against the data, as will be discussed in Chapter 10.

9.5 Empirical Variation

The theory of statistical inference allows us to estimate the likely extent of sampling errors from the single sample observed. But we usually also have comparable information from other populations or samples. This also enables us to judge the likely limits of the sampling errors in the data.

For example, in Chapter 4 we noted that the average height of certain 'privileged urban' 6-year-old boys in Ghana was 48 inches. Assuming the data was a random sample from some large population of 'privileged urban' 6-year-olds, one might want to estimate the standard error of this mean and calculate confidence limits. But no information on the sample size or on the standard deviation of the individual boys' heights was given. (Such omissions are common in practice.) Nonetheless, we can obtain a fair idea of the sampling error by looking at the data for the other boys and girls in the study shown in Table 9.2.

The means vary for the different types of 6-year-olds by about 1 inch from the overall mean of 46. The means in the other age-groups vary similarly. This variation is made up of two components:
− Real differences between the different types of children of each age,
− The sampling errors of the different samples.

The data show that neither component can vary by more than about ±1. It follows that the sampling error of the mean of the 6-year-old urban privileged boys can at most be ±1 inch or so. (Sampling errors of ±5, say, would show up as much larger variability in each column of the table.)

This kind of scrutiny of a range of similar data usually gives a good feel for one's sampling errors. In such cases we do not have to use the theoretical formula

TABLE 9.2 Average Heights of 6- to 12-Year-old Boys and Girls in Ghana

(in inches)	Age in Years 6	7	8	9	10	11	12	Av.
BOYS								
Rural	45	47	49	51	52	54	56	51
Urban	45	47	48	50	52	54	56	50
Privileged Urban	48	49	51	54	55	57	58	53
Expatriate (White)	46	49	52	54	55	58	59	53
GIRLS								
Rural	45	47	50	51	53	55	56	51
Urban	46	48	50	51	53	56	58	52
Privileged Urban	47	48	51	54	56	58	62	54
Expatriate (White)	46	49	52	53	55	59	61	54
Average	46	48	50	52	54	56	58	52

$s/\sqrt{n}$ for the standard error. The latter is merely a short-cut that allows us to assess the reliability of an isolated result, i.e. when we have no prior knowledge or other similar data.

9.6 Discussion

Estimating the population value from sample data is usually simple when dealing with basic measures like the mean or standard deviation. We just use the observed sample values as the estimates.

The likely sampling error of such a statistic depends on the unknown variability of the individual readings and their distribution shape in the population. With sample data these two features can only be estimated.

For the sample mean the solution is usually quite simple because for reasonably large samples we can assume an approximately Normal sampling distribution and ignore the fact that the standard error formula $s/\sqrt{n}$ is only an estimate. This then tells us that we are very likely to be right in supposing the population value μ to lie within about two standard error limits on either side of the observed mean m.

Such confidence limits tend to be fairly narrow unless the sample size is very small. Other sources of potential error in one's results are often much greater, e.g. whether the correct population was sampled, whether the measuring technique was appropriate, non-response errors, errors of measurement, and so on.

Problems of estimation and assessing the likely limits of sampling error do not arise for data which were not selected by a probability method, or if no explicit sampling took place. Then there is either no known connection between the sample results and the population, or the sample data *are* the population.

CHAPTER 9: GLOSSARY

Bayesian approach	Adjusting the estimates of a population parameter in the light of prior probabilities.
Bias	Any tendency for sample values to be consistently lower or higher than the population value.
Confidence level	Refers to the proportion of samples for which one would be right to say that the value of the population parameter lies within the specified confidence limits.
Confidence limits	Range of values which one can say with a given level of confidence will contain the population parameter in question.
Estimation	Using the sample data to decide on the value (or range of values) which the population parameter in question is expected to have.
Estimator	Any formula for sample data which provides an estimate of the population parameter.
Maximum likelihood	A procedure which chooses as an estimate the value which, if true, would yield the highest probability that the observed sample data would in fact have been selected.
Parameter	The numerical value of a summary measure in the population, such as the mean μ or the standard deviation σ.
Posterior probabilities	In Bayesian analysis the probabilities obtained for different possible values of the population parameter after adjusting the sample data in light of the prior probabilities.
Prior probabilities	Probabilities for the different likely values of the population parameter based on information or hunches before the new data are collected.
Statistic	An observed sample value of a summary measure, like the mean m or the standard deviation s.
Statistical inference	The process of arguing from sample results to population values.

CHAPTER 9   EXERCISES

9.1   Given the small population of four readings 0, 2, 4, 6, all possible samples of size 2, without replacement, are

$$0 \text{ \& } 2, \quad 0 \text{ \& } 4, \quad 0 \text{ \& } 6, \quad 2 \text{ \& } 4, \quad 2 \text{ \& } 6, \quad 4 \text{ \& } 6.$$

(a) Work out the mean for each of the six samples and for the population.

(b) Similarly work out the ranges.

(c) Comment.

9.2   A random sample of $n = 250$ readings gave a mean $m = 25$ and a standard deviation $s = 4$.

(a) What are the 95 percent confidence limits of the mean? What does this tell you?

(b) What can you say about the other 5 percent of samples?

9.3 (a) For a certain sample with mean $m = 12$, the 95 percent confidence limits of the mean are 10 and 14. Do you need to know the sample size to interpret the results?

(b) Which of the following possible interpretations of the confidence limits is correct?

    (i) The probability is .95 that the population mean μ lies between 10 and 14.

    (ii) For 95 percent of all such samples, the population mean μ will lie between 10 and 14.

9.4 A sample of $n = 5$ readings gave a mean $m = 25$ and a standard deviation $s = 4$.

(a) What is the t-ratio for m, if the population mean is μ?

(b) What are the 95 percent confidence limits for the population mean μ? (Use Table 8.2.)

9.5 Fifty readings taken in London of the acceleration due to gravity gave an average of $g = 23.4$ feet per second^2, with a standard deviation of 0.8. How would you take account of the prior knowledge that g near the surface of the earth is generally about 32 feet per second^2?

9.6 The following table gives the average number of purchases of ready-to-eat breakfast cereals per household over the four quarters of 1981 in five regions of the country.

	QUARTER				
	I	II	III	IV	Av.
Region					
A	9	9	10	9	9
B	7	8	9	8	8
C	7	8	9	7	8
D	7	8	8	7	7
E	6	7	8	6	6
Average	7	8	9	7	8

The data are based on random samples each quarter but the sample sizes have not been reported. What can you say about the size of the likely errors due to sampling?

CHAPTER 10

Tests of Significance

Sometimes one wants to compare an observed result with a prior hypothesis or expectation. If the observed result differs from the hypothesis but is based on a sample, a test of significance is needed. This in effect determines whether the difference is probably real or only due to sampling error.

Section 10.1 develops the general idea of tests of significance. Choosing an appropriate hypothesis to test is covered in Section 10.2. The practical importance of a statistically significant result and the various probability levels that can be used are considered in Sections 10.3 and 10.4.

The technicalities of tests of significance vary with the kind of data being considered. Detailed procedures for sample means are summarized in Section 10.5 and ones for proportions in Section 10.6.

10.1 Testing a Statistical Hypothesis

When studying an observable phenomenon we often have some prior hypothesis in mind—e.g. from previous studies, theory, or gut feeling. If we measured the whole population in question, the result would either agree with our prior hypothesis or show it to be wrong. The outcome would be clear.

But if our data were based only on a sample from that population and the sample result differed from our prior hypothesis, we would have a problem. Was our prior hypothesis wrong? Or was the sample result different only because of sampling error—i.e. was it an atypical sample?

We could determine the answer by taking a much larger sample. But with random sampling, we can avoid the extra work and cost by using a *test of significance*. This gives the probability that the difference between the sample value and the hypothesized value was only due to a sampling error. If the probability is high we accept our prior hypothesis. If it is low we reject it.

The hypothesized value is usually called the *null hypothesis*. If the test shows the difference to be highly improbable, the sample value is called 'statistically significant', i.e. the difference is probably real. There is only a small chance that the sample result would have occurred if the null hypothesis had been true. The null hypothesis is therefore rejected.

To illustrate, suppose we were measuring the rate of absenteeism in a firm of 50 000 workers. In earlier years the rate was 6 days per head, so that is our null hypothesis for this year.

If we measure the whole work-force and determine this year's rate as 8 days per head, we simply reject the null hypothesis. If instead we measure a large random sample of 5000 workers and get a mean of 8, we expect that to be pretty accurate and still reject the null hypothesis.

But if we measure a random sample of only one hundred workers and get a mean of 8, it is not obvious that this result reflects this year's true population figure. Was the latter really different from the previous year's value of 6? Maybe the 8 is only a sampling aberration? This is when we would use a test of significance.

If the null hypothesis of 6 about the population were true, the distribution of different sample means would be approximately Normal, with a mean $\mu = 6$ and an estimated standard deviation $s/\sqrt{n}$ (equal to the standard error of the mean). Here s is the standard deviation of the hundred readings in our sample. We therefore first have to calculate s from the sample. Say it is $s = 5$. It follows that the estimated standard error of the mean is $5/\sqrt{100} = 0.5$. The two-unit difference between our sample mean 8 and our hypothesized mean 6 is therefore four times the estimated standard error of the mean, i.e. $4 \times 0.5 = 2.0$.

From a table of a Normal distribution like Table 10.1, we can see that a sample value at least this different from the population mean would occur with a probability of less than .0001. Thus fewer than 1 in 10 000 samples would have a mean this different from the hypothesized value μ *if the hypothesis were true.*

TABLE 10.1   Probability Values for the Normal Sampling Distribution of the Mean
($s/\sqrt{n}$ is the estimated standard error)

Distance from the mean μ	Probability that a sample mean lies outside the stated limits
± .6 s/√n	.55 or 55%
± 1.0 s/√n	.32 " 32%
± 1.6 s/√n	.10 " 10%
± 2.0 s/√n	.05 " 5%
± 2.6 s/√n	.01 " 1%
± 3.0 s/√n	.003 " .3%
± 3.3 s/√n	.001 " .1%
± 3.9 s/√n	.0001 " .01%

We now have two choices. Do we accept that this as a highly improbable sample from a population with $\mu = 6$? Or do we reject our null hypothesis that $\mu = 6$, as not being true for this year's population?

With a probability as small as .0001 one usually rejects the null hypothesis—it was after all only a hypothesis. The difference between the sample value and the hypothesized population value is then termed *statistically significant*. It was almost certainly not caused just by a sampling error. We accept the observed sample value $m = 8$ as reflecting the true state of the population sampled, within estimated confidence limits. (At the 95 percent probability level with a standard error of .5, that puts the absentee rate within a range from 7 to 9.)

Now suppose instead we started with a hypothesis that $\mu = 8.5$. Our observed sample mean $m = 8$ with a standard error of 0.5 is just one standard error away from μ. For a Normal sampling distribution, a difference as large as that or larger would occur in about one-third of all samples, as shown in Table 10.1.

It is quite likely that our observed sample result could have occurred had the null hypothesis been true. (About one in three sample means would do so, just by chance.) This does not prove that μ really is 8.5 days per head this year. It merely shows that the sample provides no strong reason to doubt it.

10.2   The Choice of Null Hypothesis

In a test of significance one is seeking to confirm or reject the null hypothesis from the evidence of the observed sample data. The choice of hypothesis is therefore crucial. Two types can be distinguished: the 'expected' and the 'no effect' null hypotheses. These have very different implications.

The Expected Null Hypothesis

The appropriate null hypothesis to pose in a significance test is usually what the analyst expected to happen. For example, in testing a chemical additive to gasoline, the null hypothesis might be that gasoline mileage would increase by about 4 mpg, as it had before with that kind of additive.

With sample data the analyst uses a test of significance to establish whether any observed discrepancy—like 5 mpg rather than 4—can be ignored as probably merely due to an unlucky sample. If the observed difference is not significant, the analyst had predicted the correct result, 4 mpg: the observed difference is probably only due to a sampling error. In contrast, a statistically significant result would mean that things are not as the analyst expected—it is really more like 5 mpg rather than 4.

Sometimes an unexpected observation can be important. (Fleming's discovery of penicillin is a popularly quoted example.) But on the whole one wants to be right. It is not advisable to make a habit of getting unexpected (i.e. 'significant') results here. It implies that one is bad at choosing the correct hypotheses.

A danger is that it is easy to 'prove' a hypothesis chosen after the data were collected. To guard against this, the prior evidence or reasoning that led to the choice of null hypothesis needs to be reported. The ultimate test is replication of the result in subsequent studies.

The No-Effect Null Hypothesis

A common choice of null hypothesis is that the population value in question should be zero or nil—hence the name 'null hypothesis'. For example, in exploring the chemical additive to gasoline, the null hypothesis might be that the given additive should have no effect on mileage.

However, samples would generally show at least a small non-zero result even if a zero null hypothesis were true. The aim of the significance test here is to prevent the analyst from over-interpreting such a result—i.e. to claim a positive finding when there is none.

It has therefore become a tradition that a result must be 'significantly different from zero' if notice is to be taken of it. Findings tend to be labelled as significant at the 5 percent, 1 percent or .1 percent probability levels (with the symbols *, **, and *** attached as in a hotel guide). But most of the time such a test of significance is not needed. Except possibly in a first study, the analyst does not really expect *no effect*. In the case of gasoline additives, previous studies of that kind of additive would have shown an increase in mileage. The real question is how big is the effect this time, and not whether there is any effect at all. Thus a reaction against the routine use of 'no-effect' null hypotheses has developed in recent years.

In a first study it may, however, be difficult to formulate an expected null hypothesis. To guard against claiming a positive result that might only be due to sampling errors, one can set confidence limits for the observed sample result. (These were discussed in Chapter 9.) If the confidence limits include the value zero, then the real population value may be zero.

Returning to our chemical additive, suppose that the first study showed quite a large effect on mileage, but that the confidence limits include the value zero. Then it would be possible that the additive had no real effect. We would need to collect more data (or should have used a larger sample in the first place) to establish the result within narrower confidence limits, and in particular to establish more firmly whether the effect was really non-zero.

10.3   Practical Significance

Statements that a result is 'statistically significant' are sometimes over-interpreted. There are several things which the phrase does *not* mean:

(i) That the observed difference is large (only that it is probably real).
(ii) That the result is important. (The practical implications of a mileage increase depend not only on its size, but also on the cost of the additive and its side-effects, like engine wear and pollution.)

(iii) That the result will generalize (i.e. recur under other conditions, for other populations or samples from them).

Statistical significance means none of these things. It only tells us that the observed sample result most probably reflected the particular population sampled.

10.4   The Level of Significance

Assessing the statistical significance of a sample result turns on the probability that we had an unlucky sample. When we discussed simple random sampling in Chapter 7 we considered the chances of getting a highly unrepresentative sample; e.g. that all ten homes in a sample of ten would have freezers when in the population only 60 percent had them. When we consider levels of significance we are dealing with the same problem. What is the probability that the sample result observed could have come from that small percentage of atypical samples that do occur in random sampling?

The 5 percent probability level is the most commonly used cut-off point for statistical significance, i.e. for rejecting the null hypothesis μ and accepting the sample result m at its face value. With an approximately Normal sampling distribution, this probability level comes at ± 2 standard errors from μ. Such a cut-off point, while convenient, must not be applied rigidly. It makes no sense to reject a null hypothesis if m is ± 2.1 times the standard error from μ, but to accept the hypothesis if m is only ± 1.9 times the standard error from μ.

Therefore other significance levels are also used. A 10 percent level means that one declares significant any result that is more than about ± 1.5 times the standard error from the hypothesized value. Alternatively, with a 1 percent probability level a result is considered significant only if it is more than ± 2.6 times the standard error from μ. (The relationship between the probability level and the multiple of the standard error can be seen from Table 10.1.)

Different levels of significance involve us in different chances of making an error. Two kinds of error can be considered here. One is to accept the null hypothesis when it is actually false (a *Type II Error*). The other is to reject the null hypothesis when it is actually true (a *Type I Error*). The possibilities are set out in Table 10.2. We would like the chances of correct decisions to be as high as possible. But reducing the likelihood of making a Type II Error (wrongly accepting the null hypothesis) generally means *increasing* the chances of making a Type I Error (wrongly rejecting the null hypothesis). One cannot have it both ways.

With a 5 percent level of significance, we have a 5 percent chance of committing a Type I Error: rejecting the null hypothesis even when it is true. With a 1 percent significance level we reduce the chances of committing such an error. But the chance of committing Type II Errors is correspondingly increased. The null hypothesis will now be accepted as true for all results falling between ± 2 and ± 2.6 times the standard error, since we have shifted the cut-off point from two standard

**TABLE 10.2   Correct Decisions for Sample Data
and Type I and Type II Errors**

	The Null Hypothesis in the unknown Population is	
	True	False
Accept Null hypothesis	Correct Decision.	Type II Error.
Reject Null hypothesis	Type I Error.	Correct Decision.

errors to 2.6. With a 5 percent cut-off point, one would have *rejected* the null hypothesis for such cases.

On which side should the analyst err? With a new type of airplane, one would rather commit a Type I Error: send the plane back for further tests even though it might in fact be air-safe. In legal cases we would rather commit a Type II Error: let a guilty person off rather than convict an innocent one. In other cases such evaluations are more difficult to make. But in the case of statistical tests of significance the precise probability levels are usually not very important.

The reason is that most of our results are rather clear-cut. There are broadly three cases:

(i) If the sample result differs from the null hypothesis by less than about 1.5 standard errors, then there is no reason to reject the hypothesis. (At least one in ten samples would produce such a difference by chance.)

(ii) If the sample result falls outside the 2.5 standard error limits, we can reject the null hypothesis with near certainty. (Only about one in a hundred samples or less would produce such a difference just by chance.)

(iii) If the sample result falls between roughly 1.5 and 2.5 times its standard error, it comes into a 'twilight zone'. There is then a probability of between .1 and .01 (between one in ten and one in a hundred) that the result was due to an atypical sample. This is rather unlikely, but not impossible. Here the result of a test of significance is just not clear-cut.

The twilight zones are however numerically quite small. This is illustrated in Figure 10.1 for the absenteeism example. The null hypothesis is $\mu = 6$ and the estimated standard error is 0.5 for a sample of $n = 100$.

The twilight zones would only be from 4.8 to 5.2 and from 6.8 to 7.2. We would not be sure how to interpret a result falling in either of these narrow regions. If a more decisive result were needed, a larger sample should have been taken in the first place or more data need to be collected.

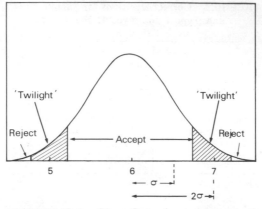

**FIGURE 10.1   Clear-Cut   Acceptance   and
Rejection Regions, and the 'Twilight Zones'**

One-Tailed Tests

In testing a null hypothesis μ, a sample mean *m* that is either much higher or much lower will lead to the hypothesis being rejected. But sometimes a 'one-tailed' cut-off point is used because an analyst may expect a deviation to occur in only *one* direction. In assessing a chemical additive to gasoline, one expects the additive either to increase the mileage or to leave it unaffected, but not to decrease it.

The difference is shown in Figures 10.2A and B. In a two-tailed test, sample means which fall into either 'tail' of the sampling distribution lead to the hypothesis being rejected. There is a 2.5 percent probability in each tail, leading to a 5 percent significance level. If the null hypothesis were true ($\mu = 20$ in the

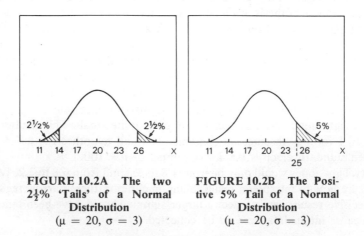

**FIGURE 10.2A   The   two
$2\frac{1}{2}$%   'Tails'   of   a   Normal
Distribution
($\mu = 20$, $\sigma = 3$)**

**FIGURE 10.2B   The   Posi-
tive 5% Tail of a Normal
Distribution
($\mu = 20$, $\sigma = 3$)**

example of Figure 10.2A), only 5 percent of the possible random samples would give means more than two standard errors above *or* below μ. Observed values like that would be regarded as unlikely, so the null hypothesis would be rejected at the 5 percent probability level.

With a *one-tailed* test, the 5 percent probability is chosen in one 'tail' only. With a Normal sampling distribution, the 5 percent cut-off point in a one-tailed test comes at 1.6 times the standard error. (This is equivalent to a 10 percent significance level for a two-tailed test, with 5 percent in the top tail and another 5 percent in the bottom tail.)

Thus if a real difference in the expected direction exists, a one-tailed test is slightly more likely to pick it up. Sample means more than 1.6 times the standard error above the hypothesized mean would be said to be significant, instead of only means more than 2.0 times the standard error. But this again puts undue emphasis on the precise probability levels. A result is *fairly unlikely* whether it occurs one in ten or one in twenty times.

In terms of Type I Errors a strict one-tailed test is however logically unacceptable. It stipulates that observed differences in the unexpected direction, however large, would never be deemed 'significant' or real. Yet such differences could occur and could be real.

Our chemical additive might certainly be expected always to increase the mileage above the normal 20 mpg. Yet a decrease to 12 mpg *could* occur, for a real and repeatable reason and not just by chance. It does not have to be due to an unlucky random sample of test drives. There might be an impurity, a mistake, a fouled-up carburetor, or whatever—something real. It would be wrong to disregard excessively low values as always being due only to chance.

So while used fairly often, one-tailed tests do not appear fully acceptable. In any case, they tend to place exaggerated importance on the precise level of significance.

10.5 Tests for Means

The formulae used in tests of statistical significance vary with the particular summary measures that are being considered. In this section we describe tests involving sample means.

The basic case of testing the difference between an observed sample mean *m* and a hypothesized population mean μ follows from the sampling distribution of the mean, as discussed in Chapter 8. For sample sizes of $n = 10$ or more, the means of different samples will generally have an approximately Normal distribution about μ with an estimated standard deviation of $s/\sqrt{n}$, where *s* is the standard deviation of the observed sample. But if the population distribution is itself highly skew, a Normal sampling distribution of the mean will arise only if *n* is greater than about 100.

The t-Test for Small Samples

For sample sizes smaller than about $n = 10$ from a population with a Normal distribution, we need to use Student's t-distribution to describe the sampling distribution of the mean. This allows for the inaccuracy of the estimated standard error $s/\sqrt{n}$.

We first calculate the standardized t-ratio

$$t = \frac{m - \mu}{s/\sqrt{n}} \, .$$

This is the difference of the observed mean m from the hypothesized population value μ, divided by the estimated standard error of m (see Section 8.5). This ratio follows Student's t-distribution with $(n - 1)$ degrees of freedom.

To determine whether the sample result supports the null hypothesis, we use a table like Table 10.3. There we find the values of t for $(n - 1)$ degrees of freedom which would be exceeded by 5 percent, 1 percent, or 0.1 percent of random samples of size n.

TABLE 10.3   Values of the *t*-distribution
(Multiples of $s/\sqrt{n}$ on either side of the mean beyond which 10%, 5%, 1% or 0.1% of the values lie)

	Degrees of freedom									
	1	2	3	4	5	10	15	20	30	Large*
10%	6	3	2.4	2.1	2.0	1.8	1.8	1.7	1.7	1.6
5%	13	4	3.2	2.8	2.6	2.2	2.1	2.1	2.0	2.0
1%	64	10	5.8	4.6	4.0	3.2	2.9	2.8	2.7	2.6
0.1%	640	32	12.9	8.6	6.9	4.6	4.1	3.8	3.6	3.3

* As for the Normal Distribution (see Table 10.1)

For example, suppose that a sample of six workers has an average absentee level of 10.2 days with a standard deviation of 4. The t-ratio, for a null-hypothesis value of $\mu = 6.0$ days is $(10.2 - 6.0)/(4/\sqrt{6})$, or about 2.6. From the table we can see that this value just reaches the 5 percent significance level for $(n - 1) = 5$ degrees of freedom. The sample result of 10.2 is therefore statistically significant at the 5 percent level. Thus even from a small random sample of six workers we can calculate that it is fairly unlikely that the absentee-rate of the population is at the hypothesized level of six days per head.

When the population distribution is not Normal (as judged from the sample, from prior knowledge, or from both), the t-ratio for small samples will not follow a t-distribution. Certain 'non-parametric' tests can then be used, as noted later in this section.

The Difference Between Two Means

A common requirement is to test the significance of the difference between the means m_x and m_y of two samples from two different populations. The difference will follow an approximately Normal sampling distribution with mean $(\mu_x - \mu_y)$ under the usual conditions (i.e. each sample size must generally be greater than about ten if the population distribution is Normal or nearly Normal, and greater than about one hundred if the population distributions are very skew). The estimated standard error is

$$\sqrt{\left(\frac{s_x^2}{n_x} + \frac{s_y^2}{n_y}\right)} .$$

The plus sign in the formula reflects that both m_x and m_y are subject to sampling error.

To illustrate, consider absenteeism results for two random samples of $n_x = 80$ and $n_y = 120$ workers from two divisions X and Y of a company. The average absentee-rates in the samples are $m_x = 8.2$ and $m_y = 7.1$ per head, and the standard deviations are $s_x = 4.2$ and $s_y = 3.8$. The null hypothesis is that absenteeism in the two divisions is the same, and in particular that $\mu_x = \mu_y$. We therefore want to assess the difference between the observed means, $m_x - m_y = 8.2 - 7.1 = 1.1$, against its standard error, $\sqrt{\{(4.2^2/80) + (3.8^2/120)\}} = 1.9$. Here the observed difference is well under one standard error and hence not significant.

For small samples, an accurate test for two sample means m_x and m_y can be derived only when the two populations are both approximately Normal and have the same standard deviation (i.e. $\sigma_x = \sigma_y = \sigma$). We calculate a t-ratio with $(n_x + n_y - 2)$ degrees of freedom:

$$t_d = \frac{(m_x - m_y) - (\mu_x - \mu_y)}{\sqrt{(s_x^2/n_x + s_y^2/n_y)}} .$$

Then we look up the appropriate values in a reference like Table 10.3. (Strictly speaking a single 'pooled' estimate of the common standard deviation σ of the two populations should be used, but the numerical effect is usually small. This is described in more advanced texts.)

Paired Readings

When comparing two samples of readings x and y, the individual readings may be *paired*. Examples are before-and-after measurements made on the same people, or pairs of patients matched by severity of illness in a clinical trial. Such pairing gives a more reliable estimate of the population difference $\mu_x - \mu_y$ if each person's measurements are relatively consistent over time.

Table 10.4 shows a small numerical example. There are big differences between the patients (e.g. patient A scoring in the 30s, and patient E only in single digits). However the before-and-after differences $d = y - x$ are fairly steady at about 4. Nonetheless, the observed differences vary from patient to patient, one having a *negative* response for example. The average difference observed could, therefore, have been due to chance. The sample might be atypical in having more patients with positive than negative differences.

TABLE 10.4   Readings for Five Patients Before and After Treatment

		Patients					Mean
		A	B	C	D	E	
Before	(x)	38	27	19	10	6	20
After	(y)	29	22	13	11	0	15
Difference (d)		9	5	6	-1	6	5

To determine whether the observed before-and-after difference was real for the population sampled, we use a test of significance based on the differences $d = (x - y)$ in the last line of the table. The mean m_d is 5 and the standard deviation s_d is

$$s_d = \sqrt{\left(\frac{\text{Sum } (d - m_d)^2}{(n - 1)} \right)}$$

$$= \sqrt{\left[\frac{(9 - 5)^2 + (5 - 5)^2 + (6 - 5)^2 + (-1 - 5)^2 + (6 - 5)^2}{(5 - 1)} \right]}$$

$$= \sqrt{(54/4)} = 3.7.$$

The estimated standard error of m_d is therefore

$$\frac{s_d}{\sqrt{n}} = \frac{3.7}{\sqrt{5}} = 1.7,$$

where n is the number of pairs of readings.

With only five readings (the five differences d) and hence $(n - 1) = 4$ degrees of freedom, we cannot use a Normal distribution. We have to calculate the t-ratio

$$t = \frac{(m_d - \mu_d)}{s_d/\sqrt{n}},$$

where μ_d is the null hypothesis value of the average before-and-after difference.

For a 'no-difference' null hypothesis where $\mu_d = \mu_x - \mu_y = 0$, the t-statistic simplifies to

$$t = \frac{m_d}{s_d/\sqrt{n}} = \frac{5}{1.7} = 2.9.$$

With four degrees of freedom this falls just above the 5 percent level of significance (see Table 10.3). We would therefore judge the observed difference $m_x - m_y = 5$ to be marginally significant.

This analysis is much more sensitive than if the data had been collected from two separate samples of five patients, i.e. five measured without treatment and five different ones with treatment. By using a 'paired' or 'matched' design, the inherent variability between patients is eliminated. Sampling errors can only arise from the variations in the before-and-after differences in each patient's scores, and not from the much larger differences between the patients' general scoring levels.

More than two Means—Analysis of Variance

Sometimes we need to compare the means of three or more samples or deal with a cross-classification between a number of patients on the one hand and more than two types of treatment on the other hand, as in Table 10.5.

TABLE 10.5   More than Two Treatments

	Patients					Average
	A	B	C	D	E	
Treatment I	38	27	19	10	6	20
" II	30	23	14	11	2	16
" III	25	20	27	3	0	11
" IV	7	5	2	0	1	3
Average	25	19	10	6	2	12

For data based on samples, tests of significance may then be needed to assess
(i) Whether the row means differ significantly from each other. (I.e. do the treatment results really differ?)

(ii) Whether the *column* means differ significantly. (I.e. do patients differ consistently from each other across all treatments?)

(iii) Whether there are significant 'interaction' effects. (Are the row differences for certain columns larger than those for other columns, meaning that some treatment effects vary from patient to patient instead of being about the same for all patients?)

Tests for differences between three or more sample means are generally performed using the Variance-ratio or F-statistic (where the F stands for Sir Ronald Fisher, who initially developed these procedures in the 1920s):

$$F = \frac{\text{Variance estimate based on sample means}}{\text{Variance estimate derived from individual readings}}.$$

This ratio follows an F-distribution whose values, depending on appropriate 'degrees of freedom', are calculated on the null hypothesis that the population means are equal. These theoretical values are published in tables given in more advanced texts. An observed value of F greater than the appropriate theoretical value indicates that at least *some* of the population means do differ at the chosen level of significance.

These procedures are an example of the *Analysis of Variance*. The method was so-named because the total variability of the data is analyzed into separate variance components, e.g. the possible differences between row means, between column means, and interaction effects.

Non-parametric Tests

We noted earlier that to use the *t*-distribution for tests of significance of the means of small samples, the distribution in the population must be approximately Normal. To deal with small samples where this assumption is not tenable, certain 'non-parametric' test procedures have been constructed. These can be used, for instance, to test whether two sets of readings, *x* and *y*, appear to come from populations with the same distributions (e.g. the same shape, the same means, etc.). The tests mostly use the *rankings* of the *x* and *y* measurements (i.e. which *x* is largest, which next largest, etc.), and not their numerical size.

For example, suppose we have the following five pairs of readings:

$$x \quad 30, \ 27, \ 11, \ 19, \ 5.$$
$$y \quad 30, \ 14, \ 23, \ 10, \ 3.$$

They could be replaced by their rank values:

Rank of *x*   1, 2, 4, 3, 5.
Rank of *y*   1, 3, 2, 4, 5.

The idea behind non-parametric tests is that assuming the 'no-difference' null hypothesis the sampling distribution of certain statistics can sometimes be theoretically calculated no matter what shape the population distribution is. The specific test procedures most widely referred to in the literature are the Mann-Whitney U-test and the Wilcoxon test. (The reader who actually needs to carry out such a test will find the details in more advanced texts.)

These test procedures are called 'non-parametric' because they involve no specification of any numerical parameters, like means or standard deviations. As a result, while one can establish whether two samples differ significantly (e.g. come from populations with different distributions), one cannot describe or estimate what the difference actually is. This is a limitation of the non-parametric approach.

10.6   Tests for Proportions

Tests of significance are sometimes required for *counts*. Here one is looking at the proportion of times some event occurs (like the numbers of days on which it rains) rather than at the mean value of a measured quantity (like the *amount* of rainfall). Tests of significance for proportions often take a technically different form than those for straight means.

For a single proportion p, the standard error for a random sample of n is given by the formula $\sqrt{\{p(1-p)/n\}}$. This is based on the short-cut version $\sqrt{\{p(1-p)\}}$ of the standard deviation of the Binomial distribution given in Chapter 5. Thus if in a random sample of 120 items from a manufacturing process, a proportion $p = .08$ or 8 percent of items were faulty, the standard error of p would be $\sqrt{(.08 \times .92/120)} = .024$, or 2.4 percent. (When p is expressed as a percentage, the standard error formula is $\sqrt{\{p(100-p)/n\}}$.)

The distribution of p in different random samples follows a Binomial distribution, as described in Section 5.3. But if the sample size is reasonably large and p is not very close to either 0 or 1, the distribution is roughly Normal. We can therefore use the simple notion of the standard error. The null hypothesis to test might be the level of defectives considered acceptable in a quality-control inspection scheme. Suppose this were $\pi = 3$ percent (Greek p, pronounced 'pi'). The difference of 5 percentage points between the sample result 8 percent and this null hypothesis is just twice the standard error of 2.4 percent. Hence the observed sample result would be significantly different from π at the 5 percent probability level. This would mean the manufacturing process was not producing acceptable batches.

More complex tests of significance involving proportions are used for contingency tables and assessments of the goodness-of-fit of a theoretical model to observed data. In both cases one uses the so-called χ^2-distribution (Greek x, pronounced 'chi' and 'chi-squared') as the required theoretical yardstick.

Contingency Tables

A contingency table is a cross-classification of qualitative data. Table 10.6 gives an example of a 2 × 2 table, i.e. two rows and two columns. (In general, one may have contingency tables of *r* rows by *c* columns.) Suppose that two machines produce the same product. From a week's output, twenty items are sampled at random from the first machine and thirty from the second. When tested for quality, ten items from the first batch and five items from the second batch are found to be defective.

TABLE 10.6   The Incidence of Defectives in Random Samples from Items Produced by Two Machines

	1st machine	2nd machine	Total
Defective	10	5	15
Non-defective	10	25	35
Total	20	30	50

If the data represented the full production that week, the first machine would be deemed worse than the second, and that would be that. But with random samples, the observed difference might only be due to sampling error.

The null hypothesis usually considered with a contingency table is that the two classifications are *independent* of each other. In our example, this means that there was no difference between the two machines during the week. The incidence of defectives would therefore not depend on which machine was used. The best estimate of the overall proportion of defectives would be 15/50 or 30 percent, from the Total column in Table 10.5. This leads to the theoretical expectation that six of the twenty items for the first machine and nine of the thirty items for the second machine should be defective, as shown in Table 10.7.

TABLE 10.7   The Observed and 'Expected' Frequencies

	1st machine		2nd machine		Total
	Obs.	Exp.	Obs.	Exp.	
Defective	10	6	5	9	15
Non-defective	10	14	25	21	35
Total	20	20	30	30	50

On the null hypothesis of independence the observed values should be equal to the expected ones, other than for sampling errors. A way of testing the significance of the differences is by calculating the sum of the quantities

$$\frac{(\text{Observed} - \text{Expected frequency})^2}{\text{Expected frequency}}$$

for each 'cell' of the table. From Table 10.7 this gives

$$\frac{(10-6)^2}{6} + \frac{(10-14)^2}{14} + \frac{(5-9)^2}{9} + \frac{(25-21)^2}{21} = 6.4.$$

The reason for performing this calculation is that under the null hypothesis of independence, the sampling distribution of the quantity

$$\text{Sum}\ \frac{(\text{Observed} - \text{Expected frequency})^2}{\text{Expected frequency}}$$

is close to a so-called χ^2-distribution, which happens to be mathematically well-documented. There are different χ^2-distributions for different sizes of contingency tables. For a table with r rows and c columns, the appropriate theoretical χ^2-distribution to look up is one with $(r-1)(c-1)$ degrees of freedom. For a 2×2 table this gives 1 degree of freedom. (The reasoning is that given the row and column totals in Table 10.6, only one figure inside the table can vary independently, e.g. the number of defectives by the first machine.)

TABLE 10.8 Values Exceeded by 5%, 1%, and 0.1% of Readings in χ^2-Distributions with various Degrees of Freedom

	Degrees of Freedom												
	1	2	3	4	5	6	8	10	12	15	20	25	30
Probability													
5%	3.8	6.0	7.8	9.5	11	13	16	18	21	25	31	38	44
1%	6.6	9.2	11.3	13.3	15	17	20	23	26	31	38	44	51
.1%	11	14	16	18	21	22	26	30	33	38	43	53	60

Table 10.8 sets out the 5 percent, 1 percent, and 0.1 percent probability values of theoretical χ^2-distributions with degrees of freedom from one to thirty. These probabilities give the percentages of samples that would have a greater χ^2-value by chance if the row and column factors in the population were an independent cross-classification.

From this table we see that the observed value of 6.4 with one degree of freedom is virtually equal to the theoretical value of 6.6 at the 1 percent probability level. Therefore a value of 6.4 would hardly ever happen by chance if the null hypothesis were true. The null hypothesis that the proportion of defectives is independent of the machines can therefore be rejected.

A technical requirement for the χ^2-distribution is that the expected number in any cell of a contingency table should be at least 5. (If it is not, some of the sub-classifications may have to be grouped.) The calculations of the quantities (Observed − Expected frequency)^2/Expected frequency must be made with the actual numbers in each cell and not with proportions or percentages, because then the theoretical distribution would not apply.

Goodness of Fit

The χ^2-distribution is also used in assessing the goodness of fit of a theoretical model, like a Normal or Poisson distribution, to sample data. Are the deviations of the observed figures from the model only due to sampling errors? By grouping the observed readings into categories with at least five readings each, the same kind of calculation as for contingency tables can be used.

To illustrate, suppose we have a sample of 200 readings with a mean $m = 9$ and a standard deviation $s = 3$, as shown in grouped form in the top line of Table 10.9. Fitting a Normal distribution with $m = 9$ and $s = 3$ gives the theoretical frequencies in the bottom line of the table.

TABLE 10.9   An Observed Distribution and a Fitted Normal Distribution

$(n = 200,\ m = 9,\ s = 3)$

	Values						Total No. of Readings
	<3	3-	6-	9-	12-	15-	
No. of readings							
Observed	7	30	61	73	25	4	200
Theoretical	5	27	68	68	27	5	200

To test whether the Normal gives a close degree of fit, we calculate the sum of the quantities {(Observed − Theoretical)^2/Theoretical}, i.e.

$$\frac{(7-5)^2}{5} + \frac{(30-27)^2}{27} + \frac{(61-68)^2}{68} + \frac{(73-68)^2}{68} + \frac{(25-27)^2}{27} + \frac{(4-5)^2}{5} = 2.6.$$

On the null-hypothesis that the population sampled follows a Normal distribution and that the deviations are merely due to sampling, the above quantity should approximately follow a χ^2-distribution with $(p - k - 1)$ degrees of freedom. Here

p is the number of groupings of readings for which observed and theoretical
frequencies are compared,

k is the number of parameters that had to be estimated in fitting the theoretical
model (in our example, the mean m and standard deviation s).

We subtract one further degree of freedom because it has been used up by fixing
the sample size n. In our example $p = 6$ and $k = 2$, so there are three degrees of
freedom.

Looking in Table 10.8 we see that the observed value of 2.6 is less than the
value of 7.8 given for the 5 percent probability level of a χ^2-distribution with three
degrees of freedom. The discrepancies from the fitted Normal distribution may
therefore be only due to sampling errors.

The value of the quantities {Sum (Observed – Theoretical)^2/Theoretical} must
always be calculated from the actual frequencies and not from percentages or
proportions. As with contingency tables, the theoretical frequency in any one
grouping interval should not be less than 5, otherwise the χ^2-distribution fails to
apply. If necessary, adjacent categories can be combined to produce this
minimum theoretical frequency, with the degrees of freedom being reduced
accordingly. In general the data can be grouped arbitrarily, but the main groupings should be decided before the data are examined in detail.

10.7 Discussion

Tests of significance are useful when sample data differ from some expected
hypothesis. Could the discrepancy just be due to sampling errors? Or is there
something real to consider? Would a much larger sample or the population as a
whole show a similar discrepancy?

Significance tests are mostly needed with small or fairly small samples, when
sampling errors can be quite large. With a large sample, virtually any difference is
real and not merely due to sampling error. But this does not mean the difference is
necessarily important.

There is little sense in testing the statistical significance of a hypothesis unless
the hypothesis is well-supported by prior knowledge. In particular, testing whether
some observed result differs significantly from zero is often pointless. It is a bit
like asking for someone who is 5 feet 8 inches tall whether this is significantly
different from zero.

If there is no well-based prior hypothesis (or if such a hypothesis has had to be
rejected because the data were significantly different), we may need to establish
confidence limits for the result, as discussed in Chapter 9. If this interval includes
zero, then it is not certain whether a real effect has been observed.

Whether we need a test of significance or merely to set confidence limits can be
judged by how we would report the results. Suppose we have some readings of the
acceleration due to gravity, with a mean of 29 feet per second^2 and a standard

error of 2.0. If we report this as 29, we need to state 95 percent confidence limits of ± 4. But if we want to report it as not significantly different from the usual value of $g = 32$ feet per second[2], we need a test of significance because we have in $g = 32$ a well-established null hypothesis.

In practice, tests of significance are mostly used when an observed difference is quite small (to see if it is 'real' or 'significant'). For large differences—large relative to the general scatter in the data—the significance is usually obvious and one does not bother to go through the technical motions of a test. But in all cases it is important to establish whether a difference also occurs in other studies and generalizes, rather than merely to test whether it really occurred on a single occasion.

CHAPTER 10   GLOSSARY

Analysis of Variance	Procedure which analyzes the total variance of a set of data by separating it into separate variance components. In effect a test of differences between mean values (see Main effect and Interaction).
Contingency table	Cross-classification of qualitative data.
Degrees of freedom	The number of readings (or with counts the number of categories) minus the number of parameters already fitted. This figure determines the particular t-, F-, or χ^2-distribution to be used.
Expected null hypothesis	Hypothesis giving the analyst's expectation of the population value.
F-statistic	Ratio between the variance estimate based on the sample means and the variance estimate derived from the individual readings in a sample. Used in the analysis of variance.
Goodness of fit	Test to determine whether a theoretical model fits sample data.
Interaction	Cases where main effects are not independent of each other (e.g. row differences varying from column to column). The term is used with the Analysis of Variance.
Main effect	In a table of rows and columns, the differences between the row means or the column means. Used with the Analysis of Variance.
No-effect null hypothesis	Hypothesis that the population value should be zero.
Null hypothesis	Hypothesized population value in a test of significance.
One-tailed test	Test of significance that assesses differences in only one direction from the hypothesized value.
Statistically significant	Term for a sample result that is different from the hypothesized value and judged not due to sampling error.
Test of significance	Determines the probability that a difference between a sample value and the hypothesized value is only due to sampling error.
Twilight zone	Area between roughly ± 1.5 and ± 2.5 standard errors, where a test of significance does not give a clear decision.
Type I Error	Accepting the null hypothesis when it is in fact false.
Type II Error	Rejecting the null hypothesis when it is in fact true.

Variance ratio	Another name for the F-statistic.
χ^2-**distribution**	Chi-squared distribution: a set of theoretical distributions for different degrees of freedom which is used in tests of significance for contingency tables and goodness of fit of theoretical models.

CHAPTER 10 EXERCISES

10.1 For a random sample of 220 books borrowed from a library in March, the average length of the loan was 12 days with a standard deviation of 6 days. The usual average length of a loan is 11 days. Is the difference significant? What does your conclusion signify?

10.2 In a random sample of 1-year-old fir trees, half had been given extra water. That half was now on average 2 inches taller than the other, with a standard error of 1.2 inches. Would you test the difference for statistical significance? Discuss what kind of null hypothesis you might use. How would you assess the practical significance of the result?

10.3 A doctor is using a random sample of patients to assess whether a new medical treatment increases the percentage recovering from an illness. Suppose he used a 1 percent probability level of significance rather than 5 percent. Would he then be more likely or less likely to say the sample result had given a significant increase in recovery rate when the treatment in fact had no effect? What would this kind of wrong conclusion be called? What other effect does the change in probability level to 1 percent have?

10.4 A small random sample of $n = 4$ had a mean of $m = 9.6$ and a standard deviation of $s = 2.2$. Is the result significantly different from a hypothesized population value of $\mu = 6$? Discuss the assumptions you are making.

10.5 Records for a random sample of 140 50-year-old men from the national census were analyzed by whether they had college educations and were divorced:

	College Education		
	Yes	No	Total
Divorced			
Yes	25	20	45
No	27	68	95
Total	52	88	140

In using a χ^2-test, what null hypothesis would you be testing? Why is there only one degree of freedom? Do you conclude that college education and divorce are associated?

PART FOUR: RELATIONSHIPS

A relationship summarizes how variables vary together. Such knowledge is central to our understanding and control of observed phenomena. In the next four chapters we examine different methods of analyzing the relationships between two or more observed variables.

One can often see a relationship by plotting the data on a graph. Either the points will be scattered with no form, implying no association, or they will at least broadly follow a definite shape, like a line or curve.

To make detailed calculations we use an equation to represent the relationship. For example, the equation $D = 40T$ states the distance D covered when traveling for T hours at a constant speed of 40 miles per hour. The numerical coefficient in the equation (here 40) gives the rate at which one variable increases for a unit increase in the other, 40 miles each hour.

If we insert different values of T into the equation, it gives the corresponding values of D. Thus $D = 40$ when $T = 1$, and $D = 120$ when $T = 3$. The values can be plotted on a graph, giving the line shown in Figure IV.1.

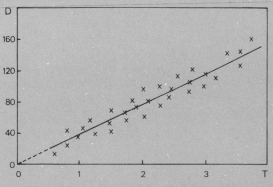

FIGURE IV.1   Distance Gone and Time Taken on
Various Journeys: Observed Values and the Line
$D = 40T$

The line would not usually fit the observed readings exactly. The measurements of actual journeys would be scattered about it. The reasons could be diverse: perhaps because of errors of measurement in D and T, perhaps because the journeys were not all at a constant rate of 40 mph. Nonetheless, the equation $D = 40T$ holds approximately within the limits of accuracy implied by the graph. (Such a result may be written $D \doteqdot 40T$; the symbol $\doteqdot$ means 'approximately equal'.)

There are three main ways of analyzing relationships between two variables. These are described in Chapters 11–13. Firstly, an index number can be computed to summarize the strength of the association between the two variables. This is the correlation coefficient, usually denoted by r. Secondly, one can calculate a regression equation which describes the form of the relationship by the line with the least amount of scatter around it in the given set of data. Thirdly, a line may be fitted to more than one set of data.

Correlation and regression analysis are the traditional methods in statistics. They are quick and easy to do and are widely employed. But both methods are applied essentially to a single set of data and the results tend to be of less general value than is often appreciated. In contrast, fitting a line to many different sets of data can give results which hold under a wider range of conditions of observation, and thus lead to the law-like relationships of science.

A variety of statistical methods have also been developed for analyzing the relationships between more than two variables. Two of the most common methods are outlined in Chapter 14: multiple regression and factor analysis.

The schema on page 155 sets out the structure of the different analytic procedures covered in the next four chapters:

Two variables   *Chapter 14: Many Variables*

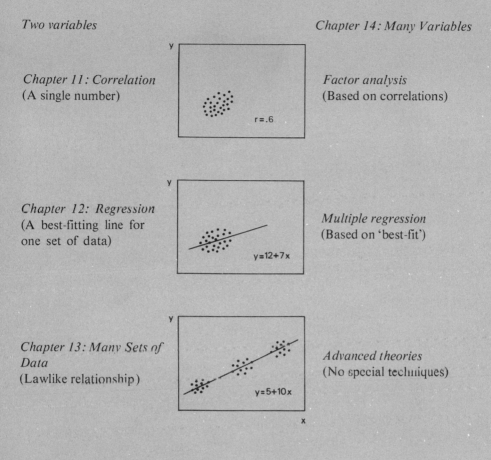

Chapter 11: Correlation   *Factor analysis*
(A single number)   (Based on correlations)

Chapter 12: Regression   *Multiple regression*
(A best-fitting line for   (Based on 'best-fit')
one set of data)

Chapter 13: Many Sets of   *Advanced theories*
Data   (No special techniques)
(Lawlike relationship)

CHAPTER 11

Correlation

The correlation coefficient is an index number which is used to indicate whether two paired variables in a given data set are associated and whether the amount of scatter in the association is large or small.

The coefficient's value ranges between −1 and +1. A correlation near +1 means little scatter: high values of y mostly occur with high values of x, and low with low. A correlation near 0 implies little or no special tendency for high y to occur with high x: sometimes they do and equally often they do not. A negative correlation arises when high y tend to go with low x and vice versa. Figure 11.1 illustrates these three cases.

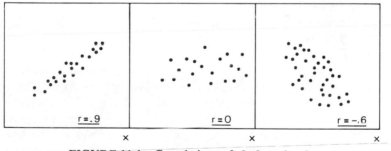

FIGURE 11.1   Correlations of .9, 0, and −.6

A general limitation of the correlation coefficient is that it does not say what the relationship actually is, i.e. how y varies with x. It only indicates the relative amount of scatter. Another limitation is that the correlation coefficient is only an appropriate measure if the relationship between two variables is roughly linear rather than curved.

11.1   The Formula

The formula for the correlation coefficient, usually denoted by r, is

$$r = \frac{\text{covariance of } x \text{ and } y}{\sqrt{(\text{variance } x)}\sqrt{(\text{variance } y)}} \, ,$$

or more briefly

$$r = \frac{\text{cov}(x, y)}{s_x s_y},$$

using the standard deviation symbols s_x and s_y.

The covariance is the crucial term here. For each pair of readings x and y, we work out the deviations from the means, $(x - \bar{x})$ and $(y - \bar{y})$, and then multiply them together as $(x - \bar{x})(y - \bar{y})$. The covariance is the average product for all pairs of readings:

$$\text{covariance of } x \text{ and } y = \frac{\text{Sum } (x - \bar{x})(y - \bar{y})}{(n - 1)}.$$

Here n is the number of pairs of readings. But as with the variance in Chapter 2, we generally use the divisor $(n - 1)$ instead of n.

The covariance summarizes how closely the two variables covary, i.e. vary together. There are four possible cases for any particular pair of readings x and y:

  (i) x is greater than the mean $\bar{x}$ and y is also greater than the mean $\bar{y}$. The deviations $(x - \bar{x})$ and $(y - \bar{y})$ are then both positive, and so is the product $(x - \bar{x})(y - \bar{y})$.

 (ii) x is smaller than $\bar{x}$ and y is also smaller than $\bar{y}$. The deviations $(x - \bar{x})$ and $(y - \bar{y})$ are then both negative, and the product $(x - \bar{x})(y - \bar{y})$ is again positive.

(iii) x is greater than $\bar{x}$ but y is smaller than $\bar{y}$. The deviation $(x - \bar{x})$ is then positive while $(y - \bar{y})$ is negative, so the product $(x - \bar{x})(y - \bar{y})$ is negative.

(iv) x is *smaller* than $\bar{x}$ but y is greater than $\bar{y}$. Here $(x - \bar{x})$ is negative, while $(y - \bar{y})$ is positive, so the product $(x - \bar{x})(y - \bar{y})$ is again negative.

In the first two cases x and y vary together—both being high or both low relative to the means—and the product of the deviations $(x - \bar{x})(y - \bar{y})$ is positive. In the second two cases x and y lie in different directions—one high and one low relative to the means—and the product $(x - \bar{x})(y - \bar{y})$ is negative.

The covariance of x and y and the correlation coefficient will be positive if the positive values of the products $(x - \bar{x})(y - \bar{y})$ predominate in the n pairs of observed readings. In contrast, if negative values of $(x - \bar{x})(y - \bar{y})$ predominate, the covariance and correlation coefficient will be negative, showing a tendency for relatively high x to go with low y, and low x with high y.

The numerical value of the covariance depends on the units of measurement of x and y. (E.g. if x is expressed in inches rather than feet, the covariance will be twelve times larger.) Dividing the covariance by the standard deviations of x and y eliminates this effect. Thus the correlation coefficient becomes an index which varies between -1 and $+1$:

$$r = \frac{\text{cov}\,(x, y)}{s_x s_y}$$

$$= \frac{\text{Sum}\,\{(x - \bar{x})(y - \bar{y})/(n - 1)\}}{\sqrt{\{\text{Sum}\,(x - \bar{x})^2/(n - 1)\}}\sqrt{\{\text{Sum}\,(y - \bar{y})^2/(n - 1)\}}}$$

$$= \frac{\text{Sum}\,(x - \bar{x})(y - \bar{y})}{\sqrt{\{\text{Sum}\,(x - \bar{x})^2\}}\sqrt{\{\text{Sum}\,(y - \bar{y})^2\}}}$$

cancelling the divisors $(n - 1)$. In the literature, r is sometimes called the 'product-moment' correlation or the Pearson correlation coefficient, after the late nineteenth-century statistician Karl Pearson.

A Numerical Example

To illustrate the calculation of the correlation coefficient with a small numerical example, Table 11.1 gives the numbers of small grocery stores (y) and large supermarkets (x) in five different districts. (E.g. there are twenty small stores and two supermarkets in District A.) The question is whether the two types of stores are related. In the example, districts with more supermarkets also have more small stores, rather than that one type of store crowds out the other. Calculating the correlation coefficient helps to summarize the pattern in the data. This can be especially useful with larger data sets.

TABLE 11.1   The Numbers of Supermarkets and Small Stores in Five Districts

	District					Mean
	A	B	C	D	E	
Supermarkets (x) :	2	3	5	7	8	5
Small Stores (y):	20	18	27	26	34	25

To calculate the correlation coefficient, we first have to write down the deviations of the five x readings from their mean $\bar{x} = 5$, and the deviations of the five ys from their mean $\bar{y} = 25$ (checking that they average at 0, as such deviations always should):

	A	B	C	D	E	
$(x - \bar{x})$	−3,	−2,	0,	2,	3.	Average 0. ✓
$(y - \bar{y})$	−5,	−7,	2,	1,	9.	Average 0. ✓

Negative deviations of x go with negative deviations of y and positive (or zero) ones with positive. This indicates already that the correlation coefficient will be positive. None of the products $(x - \bar{x})(y - \bar{y})$ are negative in this case:

	A	B	C	D	E
$(x-\bar{x})(y-\bar{y})$	15	14	0	2	27

The sum of the products is 58. To get the covariance we divide by $(n-1)$:

$$\text{cov}(x,y) = \frac{\text{Sum}(x-\bar{x})(y-\bar{y})}{(n-1)} = \frac{58}{4} = 14.5.$$

To complete the calculations for the correlation coefficient r we have to work out the standard deviations of x and y in the usual way.

$$s_x = \sqrt{\text{var}(x)} = \sqrt{\frac{\text{Sum}(x-\bar{x})^2}{(n-1)}}$$

$$= \sqrt{\frac{(9+4+0+4+9)}{4}}$$

$$= \sqrt{6.5} \doteqdot 2.5.$$

Similarly $s_y = \sqrt{\text{var}(y)} = \sqrt{40} \doteqdot 6.3$. The correlation coefficient is therefore

$$r = \frac{\text{cov}(x,y)}{s_x s_y} = \frac{14.5}{2.5 \times 6.3} = 0.90.$$

The correlation is positive, reflecting the tendency for relatively high readings of y to go with relatively high readings of x and low with low.

The numerical value of r is not 1 even though none of the products $(x-\bar{x})(y-\bar{y})$ here is negative. This reflects the fact that the observed values of x and y are not all directly proportional. For example District A has the lowest value of x but not quite the lowest value of y. When the readings are plotted, as in Figure 11.2A, the data show a linear trend but there is scatter. A correlation of 1 would occur only if all the readings lay exactly on a straight line, as in Figure 11.2B.

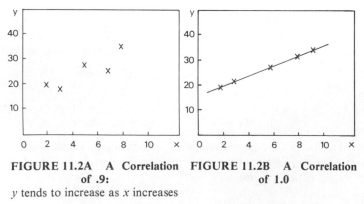

FIGURE 11.2A   A Correlation   **FIGURE 11.2B   A Correlation**
of .9:   **of 1.0**
y tends to increase as x increases

If the data had contained a pair of readings like $x = 3$, $y = 30$, with the x-value less than the mean $\bar{x} = 5$ and the y-value greater than $\bar{y} = 25$, the product of their deviations would have been negative: $(x - \bar{x})(y - \bar{y}) = (3 - 5)(30 - 25) = -2 \times 5 = -10$. The size of the covariance, and hence that of the correlation coefficient, would have been reduced.

Roughly equal numbers of positive and negative products $(x - \bar{x})(y - \bar{y})$ would lead to a low zero correlation. An excess of negative products leads to a negative correlation.

A Short-cut Formula

There is a short-cut formula for computing the covariance. It avoids having to work out all the individual deviations from the means, $(x - \bar{x})$ and $(y - \bar{y})$. This greatly speeds up the calculation, especially for large data sets, because the covariance can be computed directly from the readings. The formula is

$$\text{cov}(x, y) = \frac{\text{Sum}(xy) - n\bar{x}\bar{y}}{(n - 1)}.$$

To use this we first calculate the cross-products of the original pairs of readings x and y and sum these. For the data in Table 11.1 we quickly get

$$\text{Sum}(xy) = (2 \times 20) + (3 \times 18) + (5 \times 27) + (7 \times 26) + (8 \times 34) = 683.$$

Next we compute the term $n\bar{x}\bar{y}$, which is called the 'correction for the means':

$$n\bar{x}\bar{y} = 5 \times 5.0 \times 25.0 = 625.$$

Inserting these figures in the short-cut formula gives 14.5 as before:

$$\text{cov}(x, y) = \frac{683 - 625}{(n - 1)} = \frac{58}{4} = 14.5.$$

If we also use the corresponding short-cut formulae of Section 2.4 for the variances s_x^2 and s_y^2 (e.g. $s_x^2 = \{\text{Sum}(x^2) - n\bar{x}^2\}/(n - 1)$), the correlation coefficient can be calculated without computing any of the deviations $(x - \bar{x})$ and $(y - \bar{y})$. (To avoid undue rounding-off errors with these short-cut formulae one must be careful to use enough digits for $\bar{x}$ and $\bar{y}$.) Correlation coefficients are now often calculated on computers or programmed pocket-calculators. These generally use the short-cut formulae internally.

11.2   Sample Data

When a correlation coefficient is calculated for a random sample of paired readings from a larger population, one may wonder whether the value is significantly different from zero. Is it possible that the observed value of r was due to sampling errors and the two variables are not really correlated in the population as a whole? If the correlation *is* significant (i.e. probably real) one may want to know its standard error or confidence limits.

Statistical theory (as outlined more generally in Part Three) provides simple answers to these questions if three conditions are fulfilled. These are, (i) that the hypothesized population correlation ρ (the Greek r or 'rho') is zero, (ii) that the data are at least approximately Normal, and (iii) that the sample size n is 20 or more. If these conditions are met, the value of r in different random samples will be distributed Normally with a mean of zero and a standard error of $1/\sqrt{(n-2)}$.

To illustrate, suppose we had sample data with an observed $r = .3$ based on $n = 100$ pairs of readings. The estimated standard error of r would then be $1/\sqrt{(n-2)} = 1/\sqrt{98} = 0.1$. The observed value $r = .3$ therefore differs from the hypothesized value of $\rho = 0$ by three times its standard error. This makes the result significantly different from 0 at a probability of less than 1 percent (see Table 10.1). The best estimate of the population correlation would now be .3, with 95 percent confidence limits ranging from .1 to .5 (i.e. $.3 \pm 2 \times .1$).

It is more complicated to test a sample value of r for very small samples of n less than 20 or against a *non-zero* hypothesis for ρ. Intermediate or advanced textbooks give tests of significance for some of these situations.

11.3   Rank Correlation

Sometimes data on two variables come in the form of 'ranks' instead of quantitative measurements. This means that the n items are not measured, but rated in order from 1 to n on each variable. For example, five different types of fruit-gelatines, coded A to E, might have been ranked according to flavor and sweetness as follows:

	Gelatines				
	A	B	C	D	E
Flavor	1st	2nd	3rd	4th	5th
Sweetness	2nd	1st	4th	3rd	5th

Here low ranks on one variable tend mostly to go with low ranks on the other variable, and high with high.

To summarize this assocation, the ordinary product-moment correlation coefficient of Section 11.1 is sometimes calculated, using the ranks as measurements (despite the failure of the data to follow Normal distributions). This is then often referred to as Spearman's rank correlation coefficient ρ. The value for the gelatine example is .8.

Another correlation coefficient is Kendall's rank correlation coefficient τ (the Greek t or 'tau'). This is measured rather differently, namely in terms of the agreement of the direction of the two rankings for each *pair* of items. Thus for A&B, the directions disagree: B is ranked lower than A on Flavor (2nd versus 1st) but higher on Sweetness (1st versus 2nd). In contrast, the rankings for A&C agree in direction, A being ranked higher than C both on Flavor and on Sweetness. The numerical value of τ is given by

$$\tau = \frac{\text{No. of pairs agreeing} - \text{No. of pairs disagreeing}}{\text{Total no. of pairs}} .$$

In the gelatine example there are ten possible pairs, A&B, A&C, etc. For eight of these pairs the ranks for Flavor and those for Sweetness agree in direction, and for two they disagree (A&B and C&D). This gives $\tau = (8 - 2)/10 = .6$. (Special allowances have to be made if there are any 'tied' ranks, i.e. where two items have been ranked equal.)

Kendall's correlation coefficient τ differs numerically from Spearman's ρ. But it is again an index which can vary between -1 and $+1$. A τ of -1 indicates complete negative correlation: one set of ranks going from 1 to n while the other goes from n to 1. A τ of $+1$ means both rankings are going from 1 to n together. Zero τ implies nil correlation: no special tendency for the rankings either to agree or to disagree. The precise quantitative meaning of other values of τ, like $\tau = .6$, is not clear.

11.4   Interpreting a Correlation Coefficient

Correlation coefficients can provide a quick way of indicating whether or not two variables are related at all in the given data. If r differs from 0, the sign of the correlation also shows whether the relationship is positive or negative. The numerical size of r reflects the amount of scatter in the data. All of this can alternatively be seen by plotting the data on a graph, as in Figure 11.3. But calculating a single number like $r = -.6$ may be quicker and more convenient, especially with large data sets.

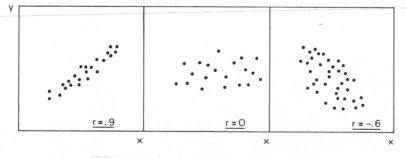

FIGURE 11.3   Correlations of .9, 0, and –.6

Trying to give the correlation coefficient a more detailed quantitative interpretation is, however, difficult. One problem is that a correlation coefficient does not describe a relationship. Saying $r = .9$, for example, does not say how much y increases for each unit increase in x. Without a graph we would not know whether the slope is flat or steep. We would only know that there is relatively little scatter about the relationship, compared with the total range of variation of x or y.

A second problem is that the numerical value of r in fact depends on two quite different things. Firstly, on how much x and y actually vary together, i.e. the numerator cov (x, y) in the formula

$$r = \frac{\text{cov}\,(x, y)}{s_x s_y}.$$

Secondly, on how much x varies on its own and how much y varies on its own, i.e. the standard deviations s_x and s_y in the denominator of r. This shows up when we compare different correlations.

Comparing Two Correlations

The interpretative problems of the correlation coefficient are highlighted when we try to compare results from different studies. Suppose two studies, one in the USA and one in the UK, both examined two variables x and y and both reported correlations of .8. Figures 11.4A and B show that the two relationships were nonetheless quite different: y increased faster in the UK than in the USA. But the relationships *might* have been the same. From the correlation coefficients alone, we could not know this.

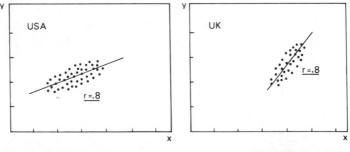

FIGURE 11.4A   US Data with   FIGURE 11.4B   UK Data with
$r = .8$ $r = .8$

Figures 11.5A and B illustrate the converse problem. Here the correlations for two studies P and Q are very different, $r = .8$ and $r = .1$. Yet from the graphs we can see that the relationship between x and y is the same. Even the scatter about the line looks the same for the two data sets.

The reason the two correlations differ so markedly is that the correlation coefficient measures the scatter about a relattionship in a *relative* way. It is

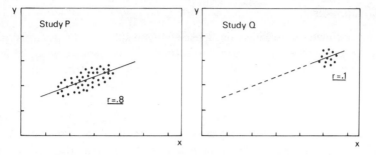

**FIGURE 11.5A   A   Correlation   FIGURE 11.5B   The Same Re-
of .8   lationship with a Correlation of .1**

measured in relation to the variation of the x and y readings themselves, s_x and s_y.
For the data set with $r = .8$, the scatter about the line is small compared to the
variation in the x and y readings. In Figure 11.5B the variation in the x and y readings is much smaller—they are much more bunched together. Compared to that
variation the scatter about the line is large, and so we get a low $r = .1$.

Thus being told merely the correlations for two variables in different studies
tells us almost nothing about how the data sets compare. All we can tell is whether
or not there is some positive or negative relationship between the two variables in
each set.

11.5   Correlation is Not Causation

There is seldom a direct causal link when two variables are correlated. This is so
even if the correlation is numerically high, like $r = .9$.

For example, there may be a marked positive correlation in a given country
between the number of murders and the number of clergymen over the last 100
years—both tending to increase. It is right to make fun of this, since there clearly
is no direct causal connection. There could be an indirect link, a 'third variable'
which caused the correlation. Both variables might have increased because the
population size increased over the years. But even this is not causally straightforward. An increase in the number of people in a country does not *necessarily*
lead to more murders and to more clergymen: a large influx of highly religious
immigrants may increase one variable but not the other. Establishing the nature of
causal mechanisms is more complex than merely noting a correlation. This will be
discussed more fully in Part Six.

A first safeguard against over-interpreting a correlation between two observed
variables is to check whether the descriptive relationship generalizes. If the same
relationship between x and y is shown to hold again and again in different data
sets, there could possibly be a causal connection. But if the descriptive relationship
differs, as for the UK and USA data in Figures 11.4A and B, there can be no
single causal connection.

11.6   Discussion

Correlation coefficients are widely used, especially in subjects like psychology and sociology. A correlation summarizes in a single index number whether or not there is assocation between two variables in the given data set. It also reflects whether the amount of scatter in the association is relatively large or small.

However, a correlation coefficient is difficult to interpret quantitatively. It does not tell us by how much one variable increases for a unit increase in the other. More particularly, the correlation coefficients in two sets of data can be numerically the same, while the relationships between x and y may be quite different. Fuller descriptions of the relationships between the variables are usually needed, as will be discussed in the next two chapters.

CHAPTER 11   GLOSSARY

Correlation	Short for correlation coefficient.
Correlation coefficient	An index number measuring the extent to which two paired variables vary together in a given data set. Often refers to the product-moment correlation.
Covariance	The average of the products of the deviations from the means for n pairs of readings x and y. $\mathrm{Cov}\,(x,\ y) = \mathrm{Sum}\,(x - \bar{x})(y - \bar{y})/(n - 1)$.
Cov $(x,\ y)$	The covariance of x and y.
Kendall's τ	A particular form of rank correlation coefficient.
Pearson correlation coefficient	A traditional name for the product-moment correlation coefficient.
Product-moment correlation	The formal name of the most widely used correlation coefficient. Defined as $r = \mathrm{cov}\,(x,\ y)/s_x s_y$. A short-cut formula is $r = [\mathrm{Sum}\,(xy) - n\bar{x}\bar{y}]/(n - 1)s_x s_y$.
r	Denotes the correlation coefficient $r = \mathrm{cov}\,(x,\ y)/s_x s_y$.
Rank	A non-quantitative form of ordering from highest (1) to smallest (n).
Rank correlation	A correlation coefficient where the two variables are the *ranks* of the n items.

CHAPTER 11   EXERCISES

11.1 (a) For the five pairs of readings

$$x \quad 0,\ 2,\ 5,\ 6,\ 7$$
$$y \quad 3,\ 7,\ 4,\ 9,\ 7$$

write out the deviations $(x - \bar{x})$ and $(y - \bar{y})$ and the products $(x - \bar{x})(y - \bar{y})$ for each pair of readings. Work out $\mathrm{Sum}\,(x - \bar{x})$, $\mathrm{Sum}\,(y - \bar{y})$, and $\mathrm{Sum}\,(x - \bar{x})(y - \bar{y})$.

(b) Calculate the covariance and the correlation coefficient from these deviations.

11.2 Use the short-cut formula of Section 11.1 to work out r for the data in Exercise 11.1.

11.3 For the five pairs of readings in Exercise 11.1, the units of x have been decreased by a factor of 10 (like going from centimetres to millimetres). This makes the x-readings 10 times as big. The data now read:

$$x \quad 0, \ 20, \ 50, \ 60, \ 70$$
$$y \quad 3, \quad 7, \quad 4, \quad 9, \quad 7$$

What effect does this have on the standard deviations of x and y, the covariance of x and y, and the correlation-coefficient?

11.4 In two studies of the variables P and Q, the following data were obtained:

First study   P   2,   3,   5,   6
                   Q   3,   2,   6,   5
Second study P   12, 13, 15, 16
                   Q   3,   2,   6,   5

Work out the correlations between P and Q and comment on the relationships between P and Q. (You may want to plot the two sets of data on a graph.)

11.5 The correlation between two variables L and M was .7 in the USA and .2 in the UK.

(a) Did L increase faster with M in the USA?
(b) Give two other reasons why the correlation in the UK may have been lower.

CHAPTER 12

Regression

Given data measuring two variables x and y, we usually want to know how the variables are related and we want to describe the relationship. One way is to plot the readings on a graph and draw a straight line or simple smooth curve through them. Figure 12.1 illustrates this with the example of the numbers of small grocery-stores (y) and supermarkets (x) from Chapter 11. Typically, there is some scatter of points about the line.

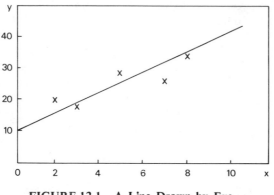

FIGURE 12.1 A Line Drawn by Eye

Since this line does not fit the observed points exactly, we could draw other lines through them, as in Figure 12.2. It is now not obvious to the eye which is the best line to pick to describe the data.

The line traditionally chosen in statistics is one that provides the 'best fit' to the data in terms of having the least scatter. This is called the *least squares regression equation*. The idea of choosing the best-fitting line is intuitively appealing, but there are problems—e.g. often there are two 'best-fitting' lines. These problems will be discussed at the end of this chapter.

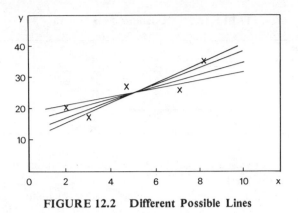

FIGURE 12.2   Different Possible Lines

12.1   A Straight-line equation

We first consider what we mean by a straight-line or 'linear' equation. It is an algebraic way of saying what a drawn line says.

To see this for the line drawn in Figure 12.1, we first note that it cuts the y-axis at $y = 10$. This tells us that when $x = 0$, the value of y is 10. This value is called the *intercept-coefficient* for the line, because that is where it 'intercepts' the y-axis.

We can determine a point at the other end of the line by using a ruler. For example when $x = 10$, the value of y on the line is about 40. Compared with the point $x - 0$, $y - 10$, y has gone up by 30 units while x has gone up by 10 units—an increase in y of $30/10 = 3$ units for each unit increase in x. This is the *slope* or *gradient* of the line. Expressed as a general formula for any two points (x_i, y_i) and (x_j, y_j) on the line

$$\text{Slope} = \frac{y_j - y_i}{x_j - x_i}.$$

We therefore know that when $x = 0$, $y = 10$ and that y increases by 3 units for every unit increase in x. As an equation this reads

$$y = 10 + 3x.$$

Having described the drawn line, we now have to assess its fit to the data.

The Fit of the Equation

The line does not say what actually happens for the two variables in our data. None of the observed points in Figure 12.1 lie exactly on the line. Nonetheless, it reflects the broad pattern in the data. To assess how well it does this we have to measure the deviations of the observed points from the line. This will allow us to

TABLE 12.1   Deviations from $y = 10 + 3x$

						Average
Observed x	2	3	5	7	8	5
Observed y	20	18	27	26	34	25
10 + 3x	16	19	25	31	34	25
y - (10 + 3x)	4	-1	2	-5	0	0

summarize the data with a simple statement like that y varies as $(10 + 3x)$ within about ± 3 units on average.

One can measure the deviations on a graph. But it is usually easier to determine the amount of scatter numerically, as shown in Table 12.1. This compares the observed y-values with the y-values given by the linear equation when we insert the observed values of x.

For the first reading of $x = 2$, the equation says that y should be $10 + (3 \times 2) = 16$. The difference between this and the observed y-value of 20 is 4. Such differences are called the deviations or residuals of the observed points from the fitted line.

The deviations $y - (10 + 3x)$ are measured in the vertical direction on Figure 12.1. Alternatively we could have measured how far the observed points are from the line *horizontally*. Or we could have plotted the data the other way round from the start, with the numbers of supermarkets on the vertical axis and the numbers of small stores on the horizonal one. This raises questions about least-squares regression to which we return in Sections 12.6 and 12.7.

To summarize the residuals in Table 12.1, we can calculate their standard deviation. This is usually termed the residual standard deviation, or rsd for short. The line in Figure 12.1 was drawn through the overall means of the data (as one would normally do to avoid systematic bias). That is why the average of the residuals in the bottom line of Table 12.1 is zero. Their standard deviation is therefore

$$\text{rsd} = \sqrt{\{\text{Sum (Residuals)}^2/(n - 1)\}},$$

using the divisor $(n - 1)$ for standard deviations with which we are familiar.

In our example this is $\sqrt{\{(16 + 1 + 4 + 25 + 0)/4\}} = 3.39$, or roughly 3.4. This says that most of the observed points lie within about 3 or 4 units on either side of the line. Thus we can summarize our data by saying that y varies as $(10 + 3x)$ within about ± 3 units on average.

Extrapolation

The equation $y = 10 + 3x$ only applies in the range of observed values, i.e. from about $x = 2$ to $x = 8$. It does not say what y should be when $x = 100$, since there are no observations at this value. In particular, the intercept-coefficient in the equation only serves to show the height of the line. It does not say that y should be 10 when $x = 0$ since we have no observed data at or near $x = 0$. If we had, the readings could for example lie closer to the zero point, as illustrated in Figure 12.3; they would therefore not lie on the line used to describe the initial data.

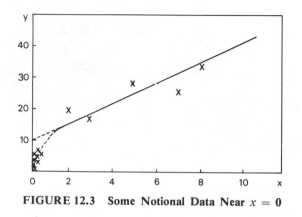

FIGURE 12.3   Some Notional Data Near $x = 0$

12.2   The Best-fitting Line

Figure 12.2 showed that the line $y = 10 + 3x$ is not the only one we could have drawn by eye for these data. Another possible line is described by $y = 12.5 + 2.5x$. Calculating the residuals for this equation as in the previous case, we find it has an rsd of 2.85. It therefore provides a closer fit than $y = 10 + 3x$ with its rsd of 3.40. Similarly, we can try other lines. We would then find by trial-and-error that the line

$$y = 14 + 2.2x$$

has a smaller residual standard deviation (2.78) than any other line. This is therefore the best-fitting line (when the residuals are measured in the vertical direction). It is called the *least squares regression equation of* y *on* x, or the regression for short. (It got the name 'least squares' because the rsd involves calculating the squared residuals.)

The trial-and-error approach would be very time-consuming in practice. There is however a theoretical short-cut which gives the regression equation in one direct calculation.

The Theoretical Regression Formula

If we write the equation of any straight line as $y = a + bx$ and let a and b also stand for the particular coefficients of the regression equation, it can be shown mathematically that the slope b of the best-fit regression equation is given by the formula

$$b = \frac{\text{covariance } (x, y)}{\text{variance } (x)}.$$

This is called the *regression coefficient*. The covariance in this formula is

$$\text{cov } (x, y) = \frac{\text{Sum } (x - \bar{x})(y - \bar{y})}{(n - 1)},$$

the same measure as in the formula for the correlation coefficient discussed in Chapter 11.

The regression coefficient can therefore be calculated as

$$b = \frac{\text{Sum } \{(x - \bar{x})(y - \bar{y})\}/(n - 1)}{\text{Sum } \{(x - \bar{x})^2\}/(n - 1)},$$

$$= \frac{\text{Sum } (x - \bar{x})(y - \bar{y})}{\text{Sum } (x - \bar{x})^2},$$

cancelling the common divisors $(n - 1)$.

Or using the short-cut versions of the variance and covariance formulae (Chapters 2 and 11),

$$b = \frac{\text{Sum } (xy) - n\bar{x}\bar{y}}{\text{Sum } (x^2) - n\bar{x}^2}.$$

The task is often simplified even further by using a computer or programmed calculator.

To determine the value of the intercept-coefficient a, we note that a line which minimizes the scatter must go through the means $\bar{x}, \bar{y}$. Hence $\bar{y} = a + b\bar{x}$, so that we can work out a from

$$a = \bar{y} - b\bar{x},$$

where b has already been calculated.

A Numerical Example

To illustrate the calculations we use the five pairs of readings in Table 12.1. The covariance works out as

$$\text{cov } (x, y) = \frac{(2 - 5)(20 - 25) + \ldots + (8 - 5)(34 - 25)}{4} = 58/4 = 14.5.$$

The variance of x is Sum $(x - \bar{x})^2/(n - 1) = \{(2 - 5)^2 + \ldots + (8 - 5)^2\}/4 = 6.5$.

The regression-coefficient is therefore

$$b = \frac{14.5}{6.5} = 2.23.$$

The intercept-coefficient is given by $a = \bar{y} - b\bar{x}$, so that

$$a = 25.0 - (2.2 \times 5.0) = 13.9.$$

This gives $y = 13.9 + 2.23x$ as the regression of y on x, or $y = 14 + 2.2x$ when rounded. It is the same equation as we found earlier by trial-and-error.

The short-cut version of the formula for the regression-coefficient gives the result more quickly:

$$b = \frac{683 - (5 \times 5.0 \times 25.0)}{151 - (5 \times 5.0^2)} = \frac{58}{26} = 2.23.$$

12.3   The Residual Scatter

The residuals from the regression line can be calculated numerically by the same method as used in Table 12.1. We can then summarize these residuals by working out their standard deviation, the rsd, in the usual way.

However, in the case of a regression equation the residual standard deviation can be calculated much more quickly by certain theoretical formulae. One version is

$$\text{rsd} = \sqrt{\{\text{var } (y) - b^2 \text{ var } (x)\}}.$$

This involves only the variances of y and x and the regression-coefficient b. We already know that var $(x) = 6.5$ and $b = 2.23$, and we can easily calculate var $(y) = 40$ from the y-readings. The rsd is therefore $\sqrt{\{40 - 2.23^2 \times 6.5\}} = 2.8$.

A popular alternative of this formula uses the correlation coefficient r of Chapter 11 and the standard deviation of y, s_y:

$$\text{rsd} = s_y\sqrt{(1 - r^2)}.$$

In our numerical example, $s_y = 6.3$ and $r = .9$ (as was calculated in Section 11.1). The rsd is therefore $6.3\sqrt{(1 - .9^2)} = 6.3 \times .44 = 2.8$ as before.

This is an important formula. It shows that the residual variation as measured by the rsd is smaller than the originally observed variation of the y-readings, as measured by s_y, by a factor of $\sqrt{(1 - r^2)}$.

To illustrate, Figure 12.4A shows the variation of the observed y-values from their mean of 25. This variation is summarized by their standard deviation $s_y = 6.3$. Figure 12.4B shows the considerably smaller scatter of the observed y-values from the regression line. These residuals are summarized by the rsd $= 2.8$.

The formula rsd $= s_y\sqrt{(1 - r^2)}$ is often said to show how much of the initial variation of y has been 'accounted for' or 'explained by' the x-variable in terms of

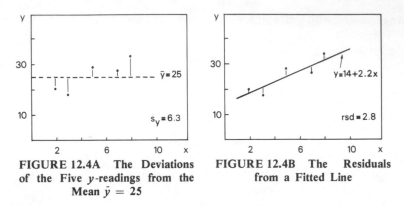

FIGURE 12.4A The Deviations of the Five y-readings from the Mean $\bar{y} = 25$

FIGURE 12.4B The Residuals from a Fitted Line

the regression equation. But such phrases can easily be over-interpreted. There is usually no direct causal connection between x and y. In our example for instance, the number of supermarkets (x) do not 'account for' or 'explain' in any direct or deep sense the number of small grocery-stores (y).

In the statistical literature, the above formula is often expressed in terms of variances. It then reads

$$\text{variance of the residuals} = s_y^2(1 - r^2).$$

These formulae provide a way of judging correlation coefficients. With a correlation of .9 as in our example, the average scatter about the fitted line (rsd = 2.8) is still almost half of the original scatter of the observed y-values ($s_y = 6.3$). For smaller correlations, even larger amounts of residual scatter are left over. Table 12.2 gives some typical values. Thus with an $r = .2$, the residual scatter is 98 percent of initially observed y-variation. In that sense, the numerical value of a correlation coefficient can give an exaggerated impression of the degree of association.

TABLE 12.2 The Relative Size of the Residual Standard Deviation for Various Values of r

($100\sqrt{(1 - r^2)}$ is the standard deviation of the residuals as a percentage of the observed standard deviation s_y)

r $100\sqrt{(1-r^2)}$		0	.2	.4	.6	.8	.9	.95	1.0
	%	100	98	92	80	60	44	31	0

The Distribution of the Residuals

A straight line or curve is said to fit the data if there are no systematic patterns in the residual deviations—otherwise a different line or curve should have been fitted. About half of the residuals should be positive and half negative, in an irregular manner. One can check this visually by plotting the data on a graph, or numerically from a tabulation of residuals, as in Table 12.1, with the readings arranged in increasing order of x or y.

As long as the rsd is smaller than s_y, the size of the residuals is less critical than the absence of systematic biases or sub-patterns. The scatter in Figure 12.5B is larger than that in 12.5A, and hence the value of r is lower. But there is still a clear association between x and y.

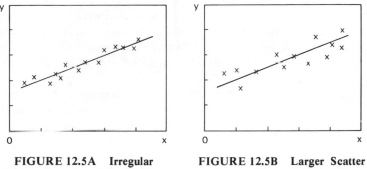

FIGURE 12.5A   Irregular   **FIGURE 12.5B   Larger Scatter**
Residuals from a Straight Line   **about a Linear Relationship**

The average size of the scatter about a fitted line is often roughly the same all along the line. The scatter is then termed *homoscedastic*. When the average size of the residuals varies greatly along the line, the condition is termed *heteroscedastic*. Summarizing the data is then more complex, as a single value of the rsd will not suffice. Figure 12.6 gives an example of each type of data.

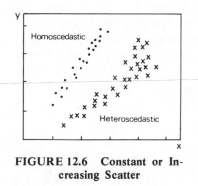

FIGURE 12.6   Constant or In-
creasing Scatter

The residuals from a well-fitting line often follow an approximately Normal distribution. This is because of the Central Limit theorem, as explained in Chapter 6. In practice, one needs to check for the data analyzed to see whether about 68 percent of the readings lie within one residual standard deviation on either side of the fitted line, and 95 percent within ± 2 rsds. If this is so, one can very simply summarize the residuals as being approximately Normal, with a stated residual standard deviation.

12.4   Sample Data

If the observed data are a random sample from a larger population, we need to estimate the coefficients of the regression equation from the sample data and to assess the accuracy of the estimates.

The Roman letters *a* and *b* are used for the coefficients of the linear regression equation ($y = a + bx$) calculated from the observed sample data (using the earlier formulae). The Greek letters α and β ('alpha' and 'beta') are used to denote the *population* values of the coefficients. To estimate these we simply use the sample values of *a* and *b*.

The estimated coefficients will be subject to sampling error. Thus the value of *b* would vary for the different samples of size *n* one might have taken. Theory shows that *b* will follow a sampling distribution with an estimated standard deviation equal to

$$\frac{\text{Residual standard deviation}}{\sqrt{\{\text{Sum }(x - \bar{x})^2\}}} .$$

This is then the standard error of the regression-coefficient *b*.

Degrees of Freedom

A complication is that in regression analysis the divisor for the residual standard deviation is usually taken to be $\sqrt{(n - 2)}$, reflecting that 2 degrees of freedom have been used up in calculating the coefficients of the regression equation (the slope and intercept-coefficients). But unless the number of readings *n* is small, the numerical difference between $\sqrt{(n - 2)}$ and $\sqrt{(n - 1)}$ is negligible. The distinction matters more in analyses involving many variables (Chapter 14).

To express the standard error formula in terms of the standard deviation of *x*, it is equivalent to either

$$\frac{\text{rsd}}{s_x\sqrt{(n - 1)}} \quad \text{or} \quad \frac{\text{rsd}}{s_x\sqrt{(n - 2)}} ,$$

depending on whether the divisor $(n - 1)$ or $(n - 2)$ has been used. Either formula shows clearly that the estimated regression-coefficient *b* is more likely to be close to the true population value of β
– The smaller the rsd (i.e. the smaller the scatter about the equation).
– The larger the spread of the observed *x*s (i.e. the larger s_x).
– The larger the sample size *n*.

In particular, if the observed sample of readings is spaced out widely (large s_x), b will be determined more precisely.

If the distribution of the residuals about the regression equation is approximately Normal, the sampling distribution of the regression estimate b is such that the ratio $(b - \beta)/$(standard error of b) follows Student's t-distribution with $(n - 2)$ degrees of freedom (see Chapters 8 and 10). If the sample size n is greater than about 20, the t-distributions are close to a Normal distribution. It then follows that the 95 percent confidence limits of β will be ± 2 times the standard error of b.

To illustrate, suppose we had a random sample of $n = 80$ pairs of readings from a larger population. Say the estimated regression equation is $y = 14 + 2.2x$ with a residual standard deviation of 2.8 and $\sqrt{\{\text{Sum } (x - \bar{x})^2\}}$ is 27. (This is equivalent to a standard deviation s_x of about 3.0.) Then the estimated slope-coefficient $b = 2.2$ would have a standard error of $2.8/27 = 0.1$. The 95 percent confidence limits for the slope would therefore be $b \pm 2 \times 0.1$, or from 2.0 to 2.4.

The standard error of b or the t-ratio may also be used in tests of significance against some specified null-hypothesis (e.g. that $\beta = 0$, or $\beta = 6$, say). Computer packages often print out the ratio of the regression-coefficient to its standard error, labelled as a 't-value'.

12.5   Non-linear Relationships

When the residuals from a straight line follow some systematic pattern, a simple curve may summarize the data better. For example, in Figure 12.7A all the negative deviations are in the middle. To summarize the residuals, one would have to spell out their pattern (positive for x up to 2, negative for x between 4 and 7, and positive again for x greater than 8). With a curve as in Figure 12.7B, the scatter would not only be smaller but it would also be irregular and hence much easier to summarize: One can simply say or imply that the residuals are irregular and of such and such an average size.

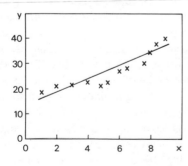

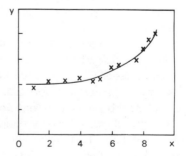

FIGURE 12.7A   Systematic Residuals from a Straight Line

FIGURE 12.7B   A Curve with Irregular Residuals

To illustrate a non-linear relationship numerically, Table 12.3 shows the number of bacteria observed in a certain study. The numbers nearly doubled every minute (roughly 10, 20, 40, 80, 160, etc.), instead of increasing by a fixed amount each minute, as they would for a straight-line relationship.

TABLE 12.3   The Number of Bacteria at Different Times

Time (in minutes) :	0	1	2	3	4	5
Number of bacteria :	10	22	39	86	156	328

A curve can be fitted by applying the least squares principle. The procedure is more complicated than for a straight line, but fitting a curve is worth doing if many different sets of data show the same general pattern. Initially the form of a suitable non-linear model has often to be guessed at, by trial and error. Sometimes one is helped by knowing something about the underlying process, e.g. that bacteria increase by splitting into new pairs. This would suggest an 'exponential growth' model.

Table 12.4 shows that such an exponential model, of the form $N = 10 \times 2^t$, fits our data well. It says that N, the number of bacteria, should double for every unit of time. Thus $N = 10$ when t is 0, $N = 20$ when t is 1, $N = 40$ when t is 2, and so on.

TABLE 12.4   The Fit of the Model $N = 10 \times 2^t$

Time t (in minutes) :	0	1	2	3	4	5
Observed number of bacteria :	10	22	39	86	156	328
The model 10×2^t :	10	20	40	80	160	320

Transformations of Scale

Curved relationships are more complex to describe than straight-line ones because y does not have a constant rate of increase with x—the slope b is not constant. But instead of plotting y against x, we can try to plot some function of y, e.g. y^2, $\sqrt{y}$, log y, $1/y$, or whatever. Or we can 'transform' x. Or both variables. In suitable cases this gives a linear relationship between the new forms of x and/or y. This makes it easier to estimate the numerical coefficients of the model, e.g. by using the earlier regression formulae for a straight line. It can also reduce problems of heteroscedasticity and non-Normality.

An example of such transformations is given by the exponential equation $N = 10 \times 2^t$ in Table 12.4. This becomes a *linear* equation if we take logarithms of both sides. It then reads log $N = \log(10 \times 2^t)$. We can write this as log $N = 1 - 0.3t$. (Looking at Table A5, log $(10 \times 2^t) = \log 10 + \log 2^t$; log $2^t = t\log 2$; log $2 = .30$, and log $10 = 1$). The equation log $N = 1 + 0.3t$ is a straight-line equation in the variables log N and t; it says that log N increases by a constant rate of 0.3 for every unit of time t.

12.6   How Good is the 'Best Fit'?

The idea of choosing a best-fitting line to describe scattered data seems like common sense. However, in practice the fit of the best-fitting line is very little better than that of many other lines one could draw. That is why it was difficult to pick a best line by eye in Figure 12.2: there was little difference in the fit of the different equations.

In our numerical example the regression $y = 14 + 2.2x$ gave the best fit, with a residual standard deviation of 2.78. But equations with slopes of 2.0 or 2.4 had residual standard deviations still as low as 2.80. The fit of the regression was only about 1 percent better, yet the slopes differed by about ± 10 percent.

Again, equations like $y = 16 + 1.8x$ and $y = 12 + 2.6x$ fitted to the same data have rsds of only about 2.9. Yet one equation says that for a unit increase in x the increase in y is 1.8 units; the other says that the rate of increase in y is 2.6 units—almost 50 percent more. (This happens even though the correlation coefficient for the data was as high as .9. For data with lower correlations, even more widely differing equations will hardly differ in the sizes of their residual scatter.)

It follows that even by its own criterion of best fit, the least squares regression approach does not differentiate very effectively between equations with markedly different slopes, e.g. between the alternative lines one might have drawn by eye in the first place. A very different approach is described in the next chapter. This chooses the line which also holds for other sets of data.

12.7   The Regressions of *y* on *x* and *x* on *y*

Another problem is that even if we decide on 'best fit' as our over-riding criterion, regression analysis does not provide a unique answer. For a given set of data, like our example of small grocery-stores and supermarkets, there will be two different least-squares regression equations. They are referred to as the regression of y on x and the regression of x on y. (The problem does not arise if one variable, x say, is experimentally controlled at certain chosen values. This is then a special case of the situation discussed in Chapter 13 where there are many different sets of data.)

So far we have only considered the regression of y on x. In our numerical example this was

$$y = 14 + 2.2x,$$

as depicted in Figure 12.8A. This gave the lowest average of the squared *vertical* distances between each point and the line.

The regression of x on y is the line in Figure 12.8B. This gives the lowest

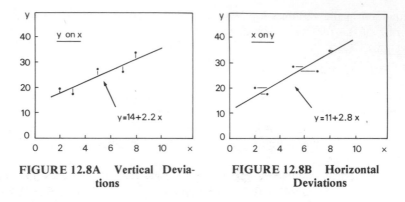

FIGURE 12.8A   Vertical Deviations

FIGURE 12.8B   Horizontal
Deviations

average of the squared *horizontal* distances between each point and the line. The equation for the regression of x on y is

$$x = -4 + .36y.$$

This says that x increases by .36 for every unit increase in y. If we transpose the equation and divide through by .36, it reads

$$y = 11 + 2.8x.$$

In this form the equation says that y increases by 2.8 units for every unit increase in x. This is about 25 percent more than for the regression of y on x (which said that y increases by 2.2 units for every unit increase in x). It is as big a difference as those between many of the equations we might have drawn by eye in the first place. With lower correlations than in our illustration (where $r = .9$), the two regression lines differ even more. Such differences persist however large the data set.

Algebraically the two equations can be written in a symmetrical form as

$$\textit{Regression of y on x} \quad (y - \bar{y}) = \frac{\text{cov}(x, y)}{\text{var } x}(x - \bar{x}),$$

$$\textit{Regression of x on y} \quad (x - \bar{x}) = \frac{\text{cov}(x, y)}{\text{var } y}(y - \bar{y}).$$

The two equations are always different unless $r = 1$, when all the observed points lie exactly on one straight line and there is no problem of choice.

Dependent and Independent Variables

To use regression analysis we must choose one of the two regression lines. To help in this, some analysts stipulate a causal direction.

For example, the sales of a product might be thought to depend on its advertising. Sales would then be called the 'dependent' variable, with advertising the 'independent' variable. This would lead one to choose the regression of sales on advertising. Yet in practice, advertising expenditure is often determined as a percentage of sales, so the question of causal direction is not solved so simply.

Our earlier example of supermarkets and small grocery-stores is typical of many other cases where there is no question of a direct causal mechanism either way. It makes no sense to ask whether x causes y or y causes x, so the choice between the two regressions remains arbitrary.

Some analysts have suggested that the choice be determined by whether one wants to predict y with least scatter from x, or x with least scatter from y. Prediction means asserting what the values of y will be in a new set of data. But successful prediction does not turn on having chosen a line that gave the best fit to the initial data set analyzed. Instead, it depends on knowing whether the result generalizes to other conditions of observation.

Minimizing the Perpendicular Deviations

A superficially attractive ideal solution would be to choose an equation by minimizing the average size of the *perpendicular* distances between the observed points and the line. But this has the crippling disadvantage that using different units of measurement for y, say, would give inconsistent answers. Thus if we were to change the units of y from yards to feet—a factor of 3—the slope-coefficient of the fitted equation would not vary by the same factor. This approach is therefore never used in practice.

12.8   Discussion

The main steps in fitting a linear regression equation to a given data set can be summarized as follows:

  (i) Decide which regression to calculate: y on x, or x on y. (There may be no simple or meaningful answer to this.)

 (ii) For the regression of y on x, calculate the regression slope b as cov (x, y)/var (x), and the intercept-coefficient a as $\bar{y} - b\bar{x}$. In our example, this gave $y = 14 + 2.2x$. The calculation is usually done on a computer or programmed calculator.

(iii) Arrange the readings in order of the size of x or y. Inspect the residuals, $y - (a + bx)$, to see if there are any systematic patterns or exceptions. Alternatively, plot the (x, y) readings and the fitted line on a graph and inspect. If the relationship looks non-linear, a *curve* will probably have to be fitted.

(iv) If the relationship is approximately linear, calculate the standard deviation of the residuals to summarize the scatter. A common short-cut formula for the rsd is $s_y\sqrt{(1 - r^2)}$.

(v) With relatively small random samples, use the standard error formula of Section 12.4 to set confidence limits or to test whether the slope-coefficient differs significantly from zero or some other prior hypothesis.

The attraction of the regression approach is that it quickly gives a 'best-fitting' line for any given data set. The method is therefore widely taught. It is extensively used in subjects like econometrics and forecasting, especially when there are more than two variables in the data, as discussed in Chapter 14. The practical value of regression analysis is however limited by three main factors:

(a) The regression equation does not provide a markedly better fit than most of the other lines that could reasonably be drawn by eye.

(b) There are in any case usually two different 'best-fit' solutions, y on x and x on y. Regression analysis therefore does not provide a unique answer. (The special case where one of the variables is controlled is discussed in the next chapter.)

(c) A regression equation fitted to one set of data (e.g. in the USA) will generally not be the best-fitting equation for any other set of data (e.g. in the UK), however large the data sets. Best-fit regression equations do not generalize.

These problems are largely resolved when we fit a relationship to two or more different sets of data, as discussed in the next chapter.

THE MAIN FORMULAE

The general equation of any straight line is $y = a + bx$ (or $y = \alpha + \beta x$ for population data), where

a the intercept-coefficient, is given either by $a = \bar{y} - b\bar{x}$ in terms of the means $\bar{x}$, $\bar{y}$ and the slope b, or if plotted on a graph, as the value of y where the line cuts the y-axis (i.e. $x = 0$).

b the slope-coefficient is given by $b = (y_j - y_i)/(x_j - x_i)$ for any two points (x_i, y_i) and (x_j, y_j) actually on the line. If one of the two points is where the line intercepts the y-axis $(x = 0)$, then $b = (y_j - a)/x_j$.

The slope of the *regression equation of* y *on* x

$$= \text{cov } (x, y)/\text{var } (x)$$
$$= \text{Sum } (x - \bar{x})(y - \bar{y})/\text{Sum } (x - \bar{x})^2,$$

or

$$= \{\text{Sum } (xy) - n\bar{x}\bar{y}\}/\{\text{Sum } (x^2) - n\bar{x}^2\}.$$

The slope of the *regression of* x *on* y

$$= \text{cov } (x, y)/\text{var } (y).$$

The covariance (x, y) in these formulae is

$$= \text{Sum } (x - \bar{x})(y - \bar{y})/(n - 1) = \{\text{Sum } (xy) - n\bar{x}\bar{y}\}/(n - 1).$$

The residual standard deviation (rsd) for the regression of y on x is

$$= \sqrt{(s_y^2 - b^2 s_x^2)},$$

or

$$= s_y \sqrt{(1 - r^2)}.$$

The standard error of the regression-coefficient b estimated from a random sample is

$$\text{rsd}/\sqrt{\text{Sum } (x - \bar{x})^2}.$$

CHAPTER 12   GLOSSARY

α	Greek alpha, the intercept-coefficient in the population.
β	Greek beta, the slope-coefficient in the population.
Deviations	Another term for residuals.
Heteroscedasticity	The average size of the residuals about a fitted line varies for different positions of the line.
Homoscedasticity	The average size of the residuals about a fitted line is roughly constant all along the line.
Intercept-coefficient (a)	The value of y where the line $y = a + bx$ cuts the y-axis (i.e. where $x = 0$). It is the sample estimate of α.
Slope (b)	The gradient of the straight line $y = a + bx$. It gives the increase in y per unit increase in x. It is the sample estimate of β.
Least-squares regression	The line that has the minimum average squared deviations.
Non-linear transformation	Expressing a variable, x or y, as a mathematical function of that variable, like y^2 or log y. Usually done to obtain a linear relationship with the other variable.
Regression of x on y	The equation taking the form $x = a + by$ which minimizes the scatter in the x or horizontal direction.
Regression of y on x	The equation taking the form $y = a + bx$ which minimizes the scatter in the y or vertical direction.
Residual standard deviation (rsd)	The standard deviation of the residuals, which is a measure of the scatter about a fitted line.
Residuals	Differences between the observed values in the data and the fitted line.
Regression coefficient	The slope b of the regression equation $y = a + bx$.
$y = a + bx$	General equation for a straight line, where a and b are either *any* numbers or the particular values given by regression analysis.
Well-fitting equation	An equation with irregular residuals (i.e. no systematic pattern in sign or size).

CHAPTER 12   EXERCISES

12.1 The means of n paired readings of x, y are $\bar{x} = 12$, $\bar{y} = 20$; the variances are var $(x) = 9$ and var $(y) = 25$; and the covariance is cov $(x, y) = 6$.

(a) What is the regression equation of y on x?

(b) Given the observation $y = 22$, $x = 6$, what is the residual from the line $y = 12 + .67x$?

12.2 (a) Calculate the regression equation of y on x for the five pairs of readings below, working in terms of the deviations $(y - \bar{y})$ and $(x - \bar{x})$:

$$x \quad 3, \quad 6, \quad 7, \quad 8, \quad 11$$
$$y \quad 13, \quad 8, \quad 11, \quad 2, \quad 6$$

(b) Now use the short-cut formula of Section 12.2 and compare your results with (a).

12.3 (a) What is the variance and the standard deviation of the residuals about the equation in Exercise 12.2? Work the values out directly from the residuals $y - (15 - x)$, and also from b or r, using the appropriate theoretical formula.

(b) How could you tell whether the residuals are irregular, homoscedastic, and approximately Normal?

12.4 For the data in Exercise 12.2, what is the regression of x on y? How does this compare with the regression of y on x? (You may want to plot the data and the two regression lines.)

CHAPTER 13

Many Sets of Data

Sometimes the same relationship holds for many different sets of data. This then leads to the empirical generalizations and law-like relationships of ordinary science, i.e. findings which hold under a wide range of different conditions and which provide the basis for the traditional predictions of science and technology.

The analysis turns on how the observed variables vary together between the different groups of readings. It is therefore sometimes called Between Group Analysis or BGA.

This type of analysis differs from the least squares regression approach of Chapter 12. There, only one set of data was analyzed and the scatter of the individual readings determined the choice of the equation to be fitted. In BGA the scatter of the individual readings within each group is a secondary issue. The emphasis is on seeing whether there is one equation which can hold for many different sets of data, at least to a close degree of approximation.

13.1 Between Group Analysis

To illustrate BGA, we start with a small set of eight pairs of readings in y and x for Germany in 1958, as shown in Figure 13.1. Here y might be the sales of different

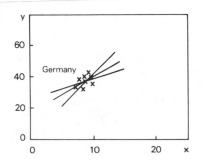

FIGURE 13.1   Various   Possible Lines for a Single Set of Readings

firms in a certain industry and x the numbers of employees. As usual, we could draw various lines through the points, some fitting better than others. In regression analysis we would choose the best-fitting line for that data and the analysis would be finished.

But with BGA the analysis is only just beginning. We now ask which line would also hold for other sets of data measuring the same variables, but under different conditions. Figure 13.2A adds one such data set: nine readings of y and x for France in the same year. There is now only one line which can hold for the means of both sets of data, as Figure 13.2B shows.

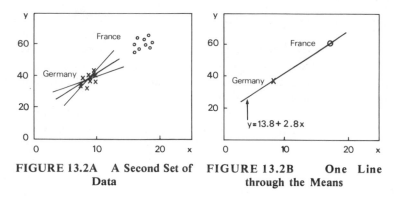

FIGURE 13.2A   A Second Set of   FIGURE 13.2B   One   Line
Data   through the Means

The line has to go through the means $(\bar{x}, \bar{y})$ of the two data sets to avoid systematic bias. Its equation is

$$y = 13.8 + 2.8x$$

Our next step is to extend the analysis to more data. Figure 13.3A shows readings for Spain, Italy and the UK as well as Germany and France, together with the earlier line. It is clear that this does not fit the data for the UK or even that for Spain. However, there is still marked correlation in the data as a whole: low ys go with low xs (Spain and the UK), and high ys go with high xs (France).

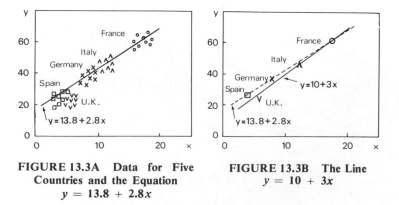

FIGURE 13.3A   Data for Five   FIGURE 13.3B   The Line
Countries and the Equation $y = 10 + 3x$
$y = 13.8 + 2.8x$

We can fit an adjusted line which will approximately describe the relationship between all five sets of readings. This is shown for the means in Figure 13.3B. The equation of the line is

$$y = 10 + 3x.$$

This equation says how y and x vary together as we go from Spain and the UK to France: as $\bar{x}$ increases from 5 to almost 20, $\bar{y}$ increases approximately as $10 + 3\bar{x}$ from 25 to 70. This occurs despite the differences between the four countries. The result is an empirical generalization, but only at a low level—it covers just five countries in one year.

The equation $y = 10 + 3x$ refers to the different means $\bar{x}, \bar{y}$ in Figure 13.3B. We have not had to consider the individual readings within each country. (We will do this in Section 13.5.) But if all the individual readings in Figure 13.3A had been pooled into a single data set, we could alternatively have tried to calculate some kind of best-fitting equation. This would however not give a single answer—we would still have to decide between the regressions of y on x and x on y.

Such a best-fit analysis of the pooled data would also not have told us explicitly that the fitted equation holds separately (if approximately) for each country. Nor would regression analysis readily accommodate yet further data sets (since regression equations are only 'best' for the data they have been fitted to). But as the range of data covered by an equation increases (other countries, different years, and so on), the extent of its generalization becomes its crucial characteristic: for predictive or interpretative uses, we are mainly concerned with the range of conditions under which the equation is already known to hold.

13.2 Fitting a Working-solution

If the means of the different data sets in Figure 13.3B had lain exactly on a straight line or a simple smooth curve, that would have been the line (or curve) to fit. There would have been no problem. But since the fitted line does not go exactly through the five pairs of means, we know there are other possible equations which could describe the data quite well. Yet these other possibilities need not greatly disturb us at this stage. The fitted line is regarded only as an early working-solution for five data sets. It is likely to be adjusted as we go on to examine an increasing range of other data sets in x and y.

To derive an initial working-solution of the form $y = a + bx$, we use the fact that the data consist of distinct groups of readings. The slope-coefficient is then calculated from the means of the two extreme sets of readings, and the intercept-coefficient by making the line go through the overall means of the data. Different analysts will obtain the same result when following this procedure, so the analysis is objective.

If the relationship between x and y across all the data sets is approximately linear, the fit will be good in that the means of the data sets will be scattered irregularly about the fitted line.

An Illustration

To illustrate we return to our example. Table 13.1 sets out the means $\bar{x}$ and $\bar{y}$ for the five European countries in 1958.

TABLE 13.1   **Means $\bar{x}$ and $\bar{y}$ for the Five European Countries in 1958**

1958	Spain	UK	Germany	Italy	France	Average
Observed $\bar{y}$	26	24	39	47	64	40
Observed $\bar{x}$	4	6	9	13	18	10

If the line is written as $y = a + bx$, we first calculate the slope-coefficient using the highest and lowest means in the formula

$$b = \frac{\bar{y}_H - \bar{y}_L}{\bar{x}_H - \bar{x}_L}.$$

The table below shows that France had the highest means in both variables, but choosing the country with the lowest means poses a problem here. Spain had the lowest $\bar{x}$ while the UK had the lowest $\bar{y}$. In such a case the means of the two data sets are combined. This gives

	France	Spain and UK
$\bar{y}$	64	25
$\bar{x}$	18	5

as the extreme means. Thus the slope-coefficient is $b = (64–25)/(18–5) = 3.0$.

Table 13.1 shows that the overall means for all five data sets were $\bar{x} = 10$ and $\bar{y} = 40$. Substituting these values into the equation $y = a + 3x$ gives a value of 10 for the intercept-coefficient:

$$40 = a + 3 \times 10$$

or

$$a = 40 - 3 \times 10 = 10.$$

The initial working-solution is therefore

$$y = 10 + 3x.$$

The Degree of Fit

We know this equation does not fit the means $\bar{x}$, $\bar{y}$ of the different countries exactly. To judge how close the approximation is we have to look at the residual deviations of the means $\bar{x}$, $\bar{y}$ from the line. This we could do on a graph like Figure 13.3B. But the numerical details are more explicit in Table 13.2.

TABLE 13.2 The Fit of the Working-solution $y = 10 + 3x$

	Spain	UK	Germany	Italy	France	Average
Observed $\bar{x}$	4	6	9	13	18	10
Observed $\bar{y}$	26	24	39	47	64	40
$10+3\bar{x}$	22	28	37	49	64	40
$\bar{y}-(10+3\bar{x})$	4	-4	2	-2	0	0*

* m.d. = 2.4

We see that the deviations are irregular in sign as x increases: positive, negative, positive, etc. We also see that the deviations vary in size: some large, some small or zero. This irregularity shows that the relationship between $\bar{x}$ and $\bar{y}$ across the five different data sets approximates a straight line rather than any simple curve. (With a curved relationship, the deviations from the fitted straight line would be consistently positive or consistently negative for at least part of the line.) The mean deviation of the residuals is 2.4.

We can therefore summarize the five sets of data by noting that the relationship between the means $\bar{x}$ and $\bar{y}$ is approximately linear, as expressed by the equation $y = 10 + 3x$, with irregular residuals having an average size of 2.4.

The Amount of Variation Accounted For

We can also look at this result in terms of the amount of $\bar{y}$-variation which it 'explains'. The $\bar{y}$-values in Table 13.1 vary from 24 for the UK to 64 for France. We could summarize this variation by saying that the $\bar{y}$-values have a mean deviation of 12.4 about their overall mean value of 40. But with the equation $y = 10 + 3x$, if we are given the $\bar{x}$ values, we can specify the $\bar{y}$-values for each country much more closely, i.e. to within an average error of only ± 2.4. Much of the $\bar{y}$-variation has been 'accounted for' or 'explained' by the variation of the $\bar{x}$s. (One must again not read too much into this kind of statement: correlation still does not mean causation.)

Had the difference between the observed $\bar{y}$-variation and the residual scatter from the fitted line been negligible, the fitted relationship would not 'account for' the $\bar{y}$-variation. There would have been no correlation between the $\bar{x}$s and $\bar{y}$s.

13.3 Prior Knowledge

The most common way of fitting a line to some given data is by using prior knowledge.

Suppose we are faced with data on the same variables x and y for the four major regions of the USA in 1968. Instead of fitting a working-solution to this new data along the lines of Section 13.2, we can simply check whether the equation for Europe in 1958, $y = 10 + 3x$, fits again.

Plotting the USA 1968 data as in Figure 13.4 shows that the earlier equation does hold again. The line is somewhat closer to the means for the north-east, south, and west than to the mid-western mean. Nevertheless, the equation fits these USA data to within the same broad limits of ± 2.4.

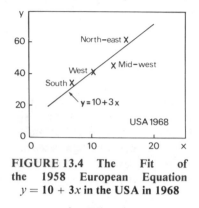

FIGURE 13.4   The   Fit   of the   1958   European   Equation $y = 10 + 3x$ in the USA in 1968

The analysis has been easy to do. What is more, it has simultaneously summarized the new data and generalized the previous result. Thus the equation $y = 10 + 3x$ is now known to hold despite differences in

– Time (1968 versus 1958),
– Place (USA versus Europe), and
– Any other background factors that varied.

These factors include differences in observers or analysts, climate, population, measuring instruments, the price of oil, interest rates, the political parties in power, and so on. Just one replication has shown that none of the differences appears to matter.

Yet the generalization is still limited, since only a few data sets were involved. The agreement between the USA and European results might even have been a coincidence, caused by two contrary factors cancelling out. This can be checked by extending the result to further data, collected under even more different conditions of observation.

Attempts to extend the range of generalization do not always succeed. For example, the 1958/68 equation $y = 10 + 3x$ might not agree with some more recent European data for 1981 which is summarized by the equation $y = 20 + 4x$ in Figure 13.5.

The failure to fit cannot merely be due to the difference in time as such, since the agreement of the 1958 and 1968 results showed that time by itself does not seem to affect the relationship. There remain a variety of other more specific causes for the failure, like the after-effects of the 1973 oil crisis, the changes in the bank rate, or even the season of the year. Maybe the relationship is generally $y = 10 + 3x$ in the summer and $y = 20 + 4x$ in the winter?

Examination of just one more set of data would begin to pin-point the possibilities. Is the new data like the 1958 and 1968 results, or like the 1981 result,

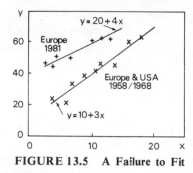

FIGURE 13.5   A Failure to Fit

or different yet again? Often such discrepancies cannot be resolved and no general pattern emerges. There may be no single linear equation or simple smooth curve for the variables x and y which will hold for all the different data sets. The data can then only be described in detail, without any simple overall generalization or summary.

13.4   Curved Relationships

Relationships between observed variables are usually curved rather than linear. Nevertheless, one often starts with linear working-solutions because they are simpler to handle and can give approximate summaries within a limited range of variation. Ambiguity over whether a straight line or curve should be fitted and the precise form of the relationship is largely resolved as one continues trying to cover data sets collected under an ever-widening range of conditions and as one also allows for theoretical considerations.

To illustrate the empirical process, we look at some data on the heights and weights of children. We start with the mean heights and weights of some 11-, 12-, and 13-year-old boys who were measured in Ghana in about 1950 (Figure 13.6A). As a first working-solution we can fit the straight line

$$w = 2.5h - 59.$$

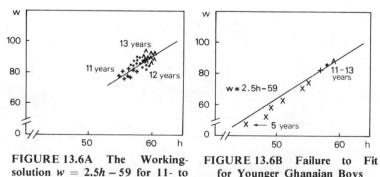

FIGURE 13.6A   The   Working-solution $w = 2.5h - 59$ for 11- to 13-year-old Ghanaian Boys

FIGURE 13.6B   Failure   to   Fit for Younger Ghanaian Boys (Age-group Means)

But this equation fails to hold when the data range is extended to younger Ghanaian boys down to the age of 5, as in Figure 13.6B. To remedy this, we can adjust the equation to fit the wider range of data. This gives the line

$$w = 3h - 91.$$

For the first three data sets (the 11- to 13-year-olds), comparison of Figures 13.7A and 13.6B shows that this new equation $w = 3h - 91$ does not fit quite as closely as the previous equation. But this is acceptable since the new equation fits a much wider range. Rather than having different equations which fit different parts of the data more precisely, it is descriptively more convenient to have a single equation which summarizes a much wider range of data to a close degree of approximation.

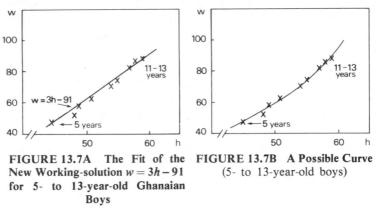

FIGURE 13.7A   The Fit of the New Working-solution $w = 3h - 91$ for 5- to 13-year-old Ghanaian Boys

FIGURE 13.7B   A Possible Curve (5- to 13-year-old boys)

However, there will be other straight line equations which can also give an approximate fit. Moreover, the deviations in Figure 13.7A are not irregular: from age 6 to 9 they are negative. This implies that we might also consider a curve, as in Figure 13.7B.

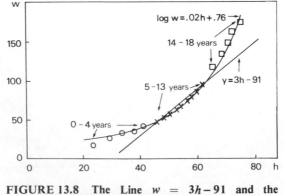

FIGURE 13.8   The Line $w = 3h - 91$ and the Curved Relationship $\log w = .02h + .76$ (0 to 18-year-olds)

The differences between such alternative solutions will not be large. But the ambiguity of choice is reduced by analyzing data that covers a still wider range, e.g. from babies of 0 to boys of 18, as shown in Figure 13.8. These readings rule out the linear working-solution $w = 3h - 91$. But the curve now clearly works well: it fits both the original and the new data. Its equation is $\log w = .02h + .76$. Only boys aged 0–1 do not lie close to this curve.

Extending the analysis to still further sets of data can lead to one of two results: either a failure to generalize, or increasing confirmation and generalization of the current working-solution. In fact, the equation above has proved to hold for a wide range of conditions. As Table 13.3 shows, they cover other countries, other races, various social and economic conditions, girls until the age of about 13, and other points in time. These include data collected 100 years ago, when children were generally smaller but, the result tells us, of much the same shape.

TABLE 13.3   Conditions Under Which the Height/Weight Relationship $\log w = .02h + .76$ Has Been Found to Hold

Race:	White, Black, Chinese (in the W. Indies).
Countries:	UK, Ghana, Katanga, West Indies, France, Canada, USA.
Time:	1880-1970 approximately.
Age:	2-18 years.
Sex:	Male (2-18) and Female (2-13).
Class:	Various in UK, France and Canada.
But NOT for:	Girls over 13. Children under 2. Undernourished children.

Three types of data do *not* fit the equation. Girls older than about 13 are generally found to be heavier than the equation says; babies are generally found to be lighter; and undernourished children are light for their height. When such exceptions generalize in quantitative detail one may want to develop different equations to summarize them too.

When further data arise they can simply be checked against the existing working-solution by examining the residual deviations. (A good qualitative check can also be made graphically.) Table 13.4 illustrates the quantitative detail with

TABLE 13.4   US Girls and the Height/Weight Relationship $\log w = .02h + .76$ (Only even years are shown)

	Age (in years)									
	0	2	4	6	8	10	12	14	16	18
Av. height $\bar h$	24	35	41	46	50	55	60	63	64	64
Av. weight $\bar w$	14	28	36	47	58	70	88	108	117	120
$\log \bar w$	1.11	1.45	1.56	1.67	1.76	1.85	1.94	2.04	2.07	2.08
$.02\bar h + .76$	1.24	1.46	1.57	1.67	1.77	1.85	1.96	2.02	0.04	2.04
$\log \bar w - (.02\bar h + .76)$	-.13	-.01	-.01	-.00	-.01	.00	-.02	.02	.03	.03

data on some USA girls. We see that, as before, the equation holds for girls between the ages of 2 and about 12, but not for babies nor older girls. (Discrepancies of .03 log weight units are not small. They amount to about 10 pounds.)

13.5   The Residual Scatter

There are two kinds of scatter about an equation that is fitted to more than one set of data: the deviations of the means $\bar{x}, \bar{y}$ of the different data sets, and the deviations of the individual readings x, y within each data set.

The deviations of the group means $\bar{x}, \bar{y}$ represent departures from the fitted relationship which are systematic for that data set. Initially these deviations may be deliberately by-passed when imposing a straight-line relationship (or simple smooth curve) onto data which are not quite that simple. But as the analysis develops one must check on the existence and generalization of any sub-patterns, as illustrated in the last section. If there are any, the fitted equation has to be adjusted or the sub-patterns have to be described separately. The average size of the deviations of the means also has to be summarized, e.g. by their standard or mean deviation.

Individual Scatter

The scatter of the individual readings in each data set has to be considered too. A basic question is whether the relationship within each group of readings is roughly of the same form as that between the different groups, as illustrated in Figure 13.9A, or quite different, as illustrated in Figure 13.9B. (The numerical analysis of individual deviations is illustrated in Section 13.6.)

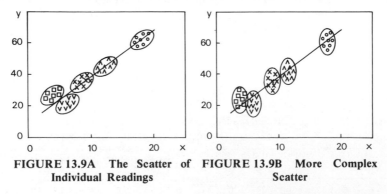

FIGURE 13.9A   The Scatter of   FIGURE 13.9B   More   Complex
Individual Readings   Scatter

The latter case raises additional questions and calls for more detailed analysis. It might be explained by there being two different relationships overall, as illustrated in Figure 13.10. Here the form of the relationship between the sales of different firms (y) and the numbers of workers (x) varies according to a third factor—the degree of automation (z). When the different firms within each country are classified by high and low degrees of automation, the relationships within each country are then parallel to the overall relationship $y = 10 + 3x$ that holds between the different countries.

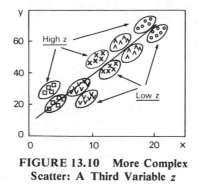

**FIGURE 13.10   More Complex
Scatter: A Third Variable z**

More generally, a crucial feature of the BGA form of analysis is that the answer does not depend on the variability of the individual readings *within* each group. It only depends on the relationship between the group means $\bar{x}, \bar{y}$. The two forms of variation—between-group and within-group—can therefore be analyzed separately. This is particularly helpful when the individual readings are not reported (as is often the case) and hence are not directly available for analysis.

One Variable Controlled

A special type of residual scatter arises when one variable is controlled at specific values. This occurs often with experimental data, where x, say, is deliberately manipulated. It also happens in observational studies when selecting particular sub-groups according to the values of one variable; e.g. measuring the heights (y) of children at selected ages (x) of 5, 9, 11, 14, etc. years.

In such cases the scatter of the individual reading within each group of readings can say nothing about how x and y are related. The only information on that is the variation *between* the groups.

Figure 13.11A illustrates a case where drug dosages of $x = 5, 9, 11, 14$, and 15 milligrams were given to different groups of patients and the responses were observed.

An equation $y = a + bx$ can be fitted to the means $(\bar{x}, \bar{y})$ of the different groups of readings as previously discussed in this chapter. The only difference is that the x-readings in each group of readings are all equal to their means $\bar{x}$. It is clear that the variation within each data set can say nothing about how x and y are related.

Usually there are some complications in this type of set-up, like errors of measurement or other forms of variability in the controlled variable x. The dosages of the drug might have been prescribed at 5, 9, 11, 14, and 15 milligrams, but the quantities actually measured out or consumed might have differed from

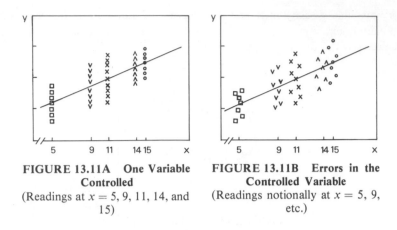

FIGURE 13.11A   One Variable Controlled
(Readings at $x = 5, 9, 11, 14$, and 15)

FIGURE 13.11B   Errors in the Controlled Variable
(Readings notionally at $x = 5, 9,$ etc.)

this, varying from case to case. (Or when children's ages are recorded to the nearest year and used as a basis for selection, the *actual* ages in each age group will vary by up to 12 months.) If the 'true' data were known by more accurate measurements, they might look as in Figure 13.11B. This is like the more general forms of data considered in previous sections in this chapter. The equation would still be fitted to the means of $(\bar{y}, \bar{x})$ of the different groups of observed readings.

13.6   Sample Data

Sometimes the different sets of data being examined are based on random samples from specific populations. Because of the nature of sampling, any particular pair of sample means $(\bar{x}, \bar{y})$ will deviate from the fitted equation $y = a + bx$ even if the equation holds almost exactly for the population data. One may then want to assess whether this difference is significant or likely to be due only to random sampling error.

To illustrate for the fitted equation $y = 10 + 3x$, suppose the 1958 data for the UK was based on pairs of readings in a random sample of $n = 7$ firms:

$$x \quad 5, \ 5, \ 6, \ 6, \ 6, \ 7, \ 7 \qquad Average \quad 6.0$$
$$y \quad 23, \ 24, \ 22, \ 24, \ 25, \ 24, \ 26 \qquad Average \quad 24.0$$

The deviation of the sample mean from the fitted equation is $24 - (10 + 3 \times 6) = -4$. This reflects that the mean for the UK lies 4 units below the fitted line. The question is whether this 4-unit difference is statistically significant. To answer this we need to determine the estimated standard error of this deviation.

First we work out the residuals from the fitted line for each pair of sample readings:

$$y - (10 + 3x): \ -2, \ -1, \ -6, \ -4, \ -3, \ -7, \ -5. \qquad Average \quad -4.$$

Then we calculate the deviations of these seven residuals from their mean of -4:

$$2, \quad 3, \quad -2, \quad 0, \quad 1, \quad -3, \quad -1. \qquad Average \quad 0.$$

This allows us to determine their standard deviation

$$s = \sqrt{\left\{\frac{\text{Sum }(4 + 9 + 4 + 0 + 1 + 9 + 1)}{6}\right\}}$$

$$s = \sqrt{\left(\frac{28}{6}\right)} = 2.2.$$

Thus the UK sample mean's difference of 4.0 from the line has an estimated error of

$$\frac{s}{\sqrt{n}} = \left(\frac{s}{\sqrt{7}}\right) = \frac{2.2}{2.6} = 0.8.$$

The difference is therefore five times bigger than its standard error. Even with a sample of $n = 7$ readings this value is well over the border-line of statistical significance. Thus we would have to conclude that the UK result reflects a real deviation in the population of UK firms that was sampled.

Such deviations from a fitted line are usually real in this way (i.e. statistically significant if based on samples). But this need not deeply affect our interpretation of the fitted equation. We do not expect it to hold exactly anyway. The justification of the fitted line remains: it is an effective summary of many different sets of data.

13.7 Discussion

When analyzing more than one set of readings the criterion of success is whether the same equation will summarize all the different data sets. The emphasis is on (i) the range of conditions under which a relationship is known to hold, and (ii) the extent to which any exceptions also generalize.

The wider and the more varied the range of conditions, the more the result tells us. For example, the relationship between children's heights and weights says that children have the same average shape despite differences in age, sex, race, country, point in time, etc.

The data required must generally consist of different groups or sub-groups of readings which explicitly relate to different conditions of observation and which also have markedly different means $(\bar{x}, \bar{y})$. When such data cannot be obtained, not enough is known about the phenomenon to establish a general relationship between x and y.

Once an initial equation has been reported and one is faced with new data, one merely has to determine whether the earlier equation holds again. This is part of the normal process of extending the equation's previous range of empirical generalization. The emphasis is therefore less on statistical techniques for fitting equations than on prior empirical knowledge, i.e. knowing results that have been reported in the past.

Thus Between group analysis deals with different questions than least-squares regression analysis. Regression chooses a best-fitting equation for a single set of readings. BGA finds an equation which holds at least approximately for many different sets of data.

CHAPTER 13   GLOSSARY

Between group analysis (BGA)	Fitting an equation to the means of different groups of readings. It describes how the means vary between the groups.
Conditions of observation	The conditions under which observations are made.
Data set	An undifferentiated group of readings.
Empirical generalization	A result which has been found to hold under a wide range of different conditions of observation.
Group of readings	Different readings that have been taken under the same conditions of observation.
Law-like relationship	A widely-established empirical generalization.
Objectivity	Establishing that different investigators get the same result.
Prior knowledge	Results known from other data.
Working solution	An equation fitted to some limited data which is likely to be adjusted in the light of further data.

CHAPTER 13   EXERCISES

13.1   The mean values of x and y for four different sets of data A to D are:

	A	B	C	D	Average
$\bar{y}$	53,	30,	12,	5	25.0
$\bar{x}$	4,	3,	1,	0	2.0

Fit two straight line equations through the overall averages, calculating the slope from
(i) The extremes,
(ii) The averages of A&B and of C&D.
What are the advantages of method (ii)?

13.2   (a) Summarize the degree of fit of the two equations in Exercise 13.1 and comment.
(b) A further set of data has means $\bar{y} = 110$, $\bar{x} = 10$. What do you conclude?

13.3   The average heights and weights of some well-off Indian boys aged 6–16 are (for alternate years):

	Age (in years)					
	6	8	10	12	14	16
Av.height $\bar{h}$ (in.)	45.7	49.6	53.5	57.5	62.2	65.7
Av.weight $\bar{w}$ (lb.)	45.8	56.1	68.9	81.8	102.5	118.8

Fit an equation to summarize the data and comment briefly. (Table A.5 gives logarithms if you wish to use these.)

13.4 Briefly state how the BGA and least-squares regression methods differ.

13.5 A random sample of $n = 6$ pairs of readings from a certain population is:

y	60,	40,	40,	30,	30,	10	Average	35.0
x	1,	2,	3,	3,	4,	5	Average	3.0

(a) The prior equation $y = 15 + 10x$ does not fit the means $\bar{y} = 35$, $\bar{x} = 3$. Is the difference significant?

(b) Comment on the nature of the scatter.

CHAPTER 14

Many Variables

Practical situations often involve relationships among more than two variables. For example, the sales of a product may vary with price levels, disposable income, the amounts of advertising, etc.

A number of statistical techniques have been developed for the analyst faced with a new set of multivariate data. Multiple regression and factor analysis are two of the more widely used techniques. They are extensions of the regression and correlation methods discussed in Chapters 11 and 12. Here we briefly outline the nature of these two techniques and note their limitations.

14.1 Multiple Regression

Multiple regression is used when there is a new data set with a number of variables and one is regarded as of particular interest—say the yield y of a crop. The analyst may want to relate the values of y to the values of the other measured variables, e.g. fertilizer x_1, temperature x_2, rainfall x_3, etc. Table 14.1 gives a small example.

TABLE 14.1   Nine Readings for Three Variables y, x_1, and x_2

										Mean
y	8.3	5.5	8.0	8.5	5.7	4.4	6.3	7.9	9.1	7.09
x_1	2.4	1.8	2.4	3.0	2.0	1.2	2.0	2.7	3.6	2.34
x_2	1.7	.9	1.6	1.9	.5	.6	1.1	1.0	.5	1.02

The first steps aim to determine whether y is related to x_1 and to x_2, and whether x_1 and x_2 are related to each other. One can see this by plotting graphs, or by calculating the correlation coefficient between each pair of variables, or by rearranging the readings in ascending order of y as in Table 14.2. Here we see that as y increases from left to right, so on the whole does x_1 and, with more exceptions, x_2. (With a larger data set we might need to group the readings and use the averages for each group.)

TABLE 14.2   **Nine Readings Arranged in Ascending Order of** y

										Mean
y	4.4	5.6	5.7	6.3	7.9	8.0	8.3	8.5	9.1	7.09
x_1	1.2	1.8	2.0	2.0	2.7	2.4	2.4	3.0	3.6	2.34
x_2	.6	.9	.5	1.1	1.0	1.6	1.7	1.3	.5	1.02

To describe these trends we can try out a linear equation of the form

$$y = a + b_1 x_1 + b_2 x_2.$$

This equation says that y should increase by b_1 units for a unit increase in x_1 when x_2 remains the same, and by b_2 units for a unit increase in x_2 when x_1 stays the same. Whether the observed data actually behave like this will have to be checked by determining how well the equation fits the data.

If the coefficients a, b_1, and b_2 are chosen by the least-squares regression principle discussed in Chapter 12, the resulting equation will have the least residual scatter. This means that the variance or standard deviation of the residuals will be less than for any other equation of that form. A residual here is the deviation or difference between the observed value of y and the value given by the equation, i.e. the difference $y - (a + b_1 x_1 + b_2 x_2)$.

The best-fitting regression equation could be found by trial and error. But it is vastly more convenient to use the theoretical formulae of Table 14.3. These allow one to calculate the coefficients a, b_1, and b_2 directly from the observed data. (The

TABLE 14.3   **The Regression Coefficients for**
$$y = a + b_1 x_1 + b_2 x_2$$

$$b_1 = \frac{r_{yx_1} - r_{yx_2} \, r_{x_1 x_2}}{(1 - r_{x_1 x_2}^2)} \times \frac{s_y}{s_{x_1}}$$

$$b_2 = \frac{r_{yx_2} - r_{yx_1} \, r_{x_1 x_2}}{(1 - r_{x_1 x_2}^2)} \times \frac{s_y}{s_{x_2}}$$

$$a = \bar{y} - b_1 \bar{x}_1 - b_2 \bar{x}_2$$

r_{yx_1} is the correlation coefficient between y and x_1;

similarly for r_{yx_2} and $r_{x_1 x_2}$.

calculations involve the standard deviations s_{y_1}, s_{x_1}, and s_{x_2} of the three variables and the pair-wise correlation coefficients r_{yx_1}, r_{yx_2}, and $r_{x_1x_2}$. Nowadays the detailed work is usually done on a computer. If the correlation r_{yx_2} is zero, the formula boils down to the simple least-squares regression equation between y and x_1 discussed in Chapter 12.)

For our numerical example the formulae give the equation

$$y = 1.1 + 2.0x_1 + 1.3x_2.$$

Mathematically this says that y should vary by 2 units for a unit change in x_1 and by 1.3 units for a unit change in x_2. In particular, this equation says that the rate at which y varies with x_1 does not depend on the values of x_2: the slope-coefficient for x_1 does not change when x_2 changes values. The unresolved question is whether the observed data really behaved in this way; if not, a more complex model has to be fitted.

The Fit of the Equation

The fit of the equation has to be assessed in two respects:
  (i)   The size of the residuals $y - (1.1 + 2.0x_1 + 1.3x_2)$,
  (ii) Whether the residuals show any systematic patterns (implying that we have fitted the wrong form of equation).

Table 14.4 illustrates the calculations. The residuals here appear irregular: some are positive and some negative, in no particular order. The use of a linear equation therefore seems appropriate. (Whether the data also conform with the notion of independence—i.e. that the effect of x_1 does not vary with the value of x_2—also needs to be checked. But this is difficult to do with only a small set of readings.)

TABLE 14.4   Residuals from the Regression Equation
$$y = 1.1 + 2.0x_1 + 1.3x_2$$

		Std.Dev.
Observed y	4.4 5.6 5.7 6.3 7.9 8.0 8.3 8.5 9.1	1.6
$1.1 + 2.0x_1 + 1.3x_2$	4.3 5.9 5.8 6.5 7.8 8.0 8.1 8.8 8.9	1.6
$y - (1.1 + 2.0x_1 + 1.3x_2)$	.1 -.3 -.1 -.2 .1 0 .2 -.3 .2	.2

The standard deviation of the residuals is .22. This is much smaller than the standard deviation of the observed values of y, which is 1.6. A large part of the observed y-variation has therefore been 'accounted for' or 'explained by' the fitted relationship with x_1 and x_2. (Such phrases are widely used but, as noted in Chapter 12, they may wrongly hint at causality.)

The fit of multiple regression equations is often reported in terms of the multiple correlation coefficient R. This equals the correlation between the observed y-values and the y-values given by the equation. (It is called a *multiple* correlation coefficient because it measures the association between the ys and all the different *x*-variables together.) It can be calculated in the same way as *r* in Chapter 11. In our example, we would calculate the correlation from the nine pairs of figures in the first two rows of Table 14.4. The value of *R* is .99. This is high and reflects that the *y*-values given by the equation are close to the observed *y*-values.

There is also a theoretical formula for *R* which depends only on the correlation between each pair of variables. It avoids having to compute the 'theoretical' *y*-values (as given by the equation in the second line of the table). In practice neither method of calculation has to be used much nowadays since the value of *R* is usually given automatically as part of the output for any multiple regression computer package.

As with simple regression in Chapter 12, *R* can be used to relate the standard deviation of the residuals to the standard deviation s_y of the observed values of *y*. The formula is

$$\text{Standard deviation of residuals} = s_y\sqrt{(1 - R^2)}.$$

The higher the value of *R*, the smaller the residual scatter will be compared with s_y.

To illustrate, in our example with *R* = .99 the formula says the standard deviation of the residuals should be equal to s_y times $\sqrt{(1 - .99^2)}$, or $1.6 \times .14 = .22$. This agrees with the value we calculated directly from the last line of Table 14.4. But the formula is quicker to use because s_y and *R* can be calculated without working out the individual residuals.

Often the calculations are expressed in terms of R^2, the square of the multiple correlation coefficient. This then represents the proportion of the *variance* of *y* that is 'accounted for' by the regression equation. Thus

$$\text{variance of residuals} = s_y^2(1 - R^2),$$

or

$$\frac{\text{variance of residuals}}{\text{variance of } y} = 1 - R^2.$$

In our example *R* is .99 and $R^2 = .98$. Thus 98 percent of the total *y*-variance has been 'accounted for' by the fitted equation. This is a high value. However, use of the variance effectively clouds the issue, since the standard deviation is the more descriptive measure of scatter.

The residual *standard deviation* is still 14 percent of s_y, since $\sqrt{(1 - .99^2)} = .14$. So the amount of scatter left in the data after the multiple regression equation has been fitted is not negligible. (When an equation is fitted to data with an *R* of .5, say, the residual scatter is almost 90 percent of the original *y*-variation, since $\sqrt{(1 - .5^2)} = .87$.)

If the number of readings is small (as in our example here), special allowance should be made for the 'degrees of freedom' when calculating the residual standard deviation. The number of parameters estimated in the regression equation has to be allowed for, as discussed in advanced texts. This leads to an 'adjusted' multiple correlation coefficient. Most computer packages print out the adjusted values of R or R^2. But the numerical effects of such adjustments are generally small.

Statistical Significance

If the data are a random sample from a specified larger population, the standard errors of the estimated slope-coefficients b_1, b_2, etc. can be used to assess the effects of the sampling. With a small sample, the numerical values of the regression coefficients can be subject to marked sampling errors and this has to be allowed for in any interpretation of the equation. (The unknown population value of the regression coefficients are denoted by β's, i.e. Greek b's, call 'betas'.)

The values of the standard errors of the regression coefficients are generally given in computer print-outs, often in the form of t-ratios (i.e. the regression coefficient divided by its standard error) or a related F-statistic ($F = t^2$). The appropriate degrees of freedom are $(n - k - 1)$, where n is the sample size and k is the number of slope-coefficients fitted. (In our example $n = 9$ and $k = 2$.)

When the degrees of freedom are less than about 10, the t-distribution has to be used to judge significance. Otherwise one uses the Normal distribution, as discussed in Part Three. A regression coefficient which is less than about twice its standard error, or one which has a t-ratio of 2 or less, may then be regarded as not significantly different from zero.

If formal tests of significance are not used, care must be taken that the number of x-variables being fitted is substantially less than the total number of readings.

14.2　Problems with Multiple Regression

Multiple regression is easy to do with a computer and is used in subjects like econometrics, statistical forecasting, and the analysis of time-series. But there are problems.

As with simple regression in Chapter 12, the answer depends on which way the regression is taken: regressing y on x_1 and x_2, as in our example, or regressing x_1 on y and x_2, etc. The xs are often called the 'independent' variables and y the 'dependent' one, but this terminology evades the real question of what depends on what. Do sales depend on price levels and advertising budgets, or vice versa?

Another problem is that equations with markedly different slope-coefficients can give almost as good a fit (as high an R^2) as the multiple regression equation itself. Thus it is not necessarily the definitive relationship for the variables.

Statistical writers also warn very strongly against interpreting the regression coefficients (b_1, b_2, etc.) as showing how much each x-variable affects y: correlation is not causation. But such warnings are widely ignored in practice. With an equation like $y = 1.1 + 2.0x_1 + 1.3x_2$ it is too tempting to suppose that a unit increase in x_2 in real life will on average lead to (or 'cause') a 1.3 unit increase in y. An additional complication is that in practice, changes in x_2 could tend to be correlated with changes in x_1.

The Choice of Variables

A special problem is the choice of variables to use in the equation. It is common to start with data on quite a number of different x-variables, say 10 or 20, and to prune them by selecting the few which account for most of the observed variation in the y-variable.

The selection procedure is carried out by first calculating the correlations of y with each of the xs. If the highest of these correlations is with x_5 say, one works out the simple regression of y on x_5. Next one calculates the residuals $\{y - (a + bx_5)\}$ from this equation and sees in effect which of the remaining x-variables is most highly correlated with these residuals. Say x_2 is the one. Then the multiple regression of y on both x_5 and x_2 is calculated. The third x-variable selected is similarly the one most highly correlated with the residuals from *this* equation, and so on. One stops adding x-variables to the equation when doing so produces little or no reduction in the residual standard deviation (or equivalently little or no increase in the multiple correlation R). This kind of selection procedure (somewhat oversimplified here) is referred to as 'step-wise regression'. It is fully automated by suitable computer programs.

With this procedure it is possible for an x-variable not to be selected for the regression equation even though it is quite highly correlated with the y-variable. This occurs if that x-variable is highly correlated with another x that has already been used. Table 14.5 gives an example by adding a third variable, x_3, to the previous illustration.

TABLE 14.5   Nine Readings on Variables y, x_1, x_2, and x_3

										Mean
y	4.4	5.6	5.7	6.3	7.9	8.0	8.3	8.5	9.1	7.1
x_1	1.2	1.8	2.0	2.0	2.7	2.4	2.4	3.0	3.6	2.3
x_2	.6	.9	.5	1.1	1.0	1.6	1.7	1.3	.5	1.0
x_3	.4	1.2	.5	1.0	1.3	2.0	3.3	.9	.5	1.2

Variable x_3 has a .40 correlation with y. But including it in the equation only reduces the residual standard deviation from .22 to .18. This occurs because x_3 has an almost .9 correlation with x_2. One variable therefore accounts for much the same variation as the other. Since x_2 was already chosen for the regression equation (having a higher correlation with y than x_3 had), there is no real need in the equation for x_3. Thus it would usually not be used.

This situation is known as *collinearity*: x-variables which vary (linearly) together. It can lead to the almost accidental selection of one variable rather than another for the multiple regression equation. In terms of 'accounting for the observed variation of y' this choice makes little difference. Using x_2 rather than x_3 gives much the same numerical answer. But any attempt at *causal* interpretation is then made impossible. For example, if x_2 is the average temperature during the growing season and x_3 is the amount of rainfall, both variables should be considered when studying crop yield y. It would be wrong to conclude that x_2 matters and x_3 does not.

14.3   Factor Analysis

Factor and component analysis are techniques that are sometimes used when there is a single data set with a large number of variables but no variable is of special interest. For example, suppose we have thirty different body measurements for a given sample of people: height H, weight W, leg length L, arm length A, girth G, chest circumference C, and so on. If the aim is to reduce the number of variables to fewer underlying concepts, perhaps like 'Size' and 'Shape', factor or component analysis could be used.

The reduction is done by creating new variables (called factors or components) which largely account for the variation in the given data, i.e. which correlate highly with the observed variables. A component is simply a linear combination or weighted average of all the observed variables:

$$\text{Component} = aH + bW + cL + dA + eG + fC + \text{etc.}$$

A factor is a more complex version of the same equation with an 'error' term added. The a, b, c, etc. are numerical coefficients. These are determined by the particular version of component or factor analysis that is adopted. The methods are fairly widely used but involve a good deal of somewhat arbitrary judgment and are often regarded as of doubtful validity.

For many purposes no major distinction need be made between factors and components; the two terms are sometimes used interchangeably. Factor analysis was developed earlier in this century in the analysis of intelligence tests in psychology, so the observed variables are often called 'test variables'.

The usual starting-point for both factor and component analysis is a *correlation matrix*. This is a square array of the correlation coefficients between all pairs of variables, as shown in Table 14.6. Here r_{WH} and r_{HW} represent the same thing: the correlation between weight W and height H.

Large correlation matrices are often regarded as too complex to analyze by inspection (although usually more could be done than is attempted). The procedures of factor or component analysis are then sometimes used instead. They differ from multiple regression because the newly created components or factor variables are not directly observed. Having created these new variables, the analyst then has

TABLE 14.6 A Correlation Matrix

	H	W	G	L	etc
H	1	r_{WH}	r_{GH}	r_{LH}	·
W	r_{HW}	1	r_{GW}	r_{LW}	·
G	r_{HG}	r_{WG}	1	r_{LG}	·
L	r_{HL}	r_{WL}	r_{GL}	1	·
etc	·	·	·	·	1

still to establish what they might mean. This is usually done simply by inspection, i.e. by seeing which of the original variables correlate highly with the new factors. The factors are then given names accordingly.

The calculations are nowadays computerized. There is a large range of procedures to choose from and the results can differ both quantitatively and qualitatively, involving arbitrary choices. But the following example covers much that is common to the more popular applications.

A Numerical Illustration

Table 14.7 gives the correlations between ten different variables which measure the extent to which a sample of adults in the UK said they 'really like to watch' each of ten different television programs, five on sports and five on current affairs. The correlation between liking Professional Boxing and liking This Week is low (0.1064); that between Professional Boxing and World of Sport is higher (0.5054).

TABLE 14.7 Correlations between Ten TV Programs

	PrB	ThW	Tod	WoS	GrS	LnU	MoD	Pan	RgS	24H
Professional Boxing	1.0000	0.1064	0.0653	0.5054	0.4741	0.0915	0.4732	0.1681	0.3091	0.1242
This Week	0.1064	1.0000	0.2701	0.1424	0.1321	0.1885	0.0815	0.3520	0.0637	0.3946
Today	0.0653	0.2701	1.0000	0.0926	0.0704	0.1546	0.0392	0.2004	0.0512	0.2432
World of Sport	0.5054	0.1424	0.0926	1.0000	0.6217	0.0785	0.5806	0.1867	0.2963	0.1403
Grandstand	0.4741	0.1321	0.0704	0.6217	1.0000	0.0849	0.5932	0.1813	0.3412	0.1420
Line-up	0.0915	0.1885	0.1546	0.0785	0.0849	1.0000	0.0487	0.1973	0.0969	0.2661
Match of the Day	0.4732	0.0815	0.0392	0.5806	0.5932	0.0487	1.0000	0.1314	0.3261	0.1221
Panorama	0.1681	0.3520	0.2004	0.1867	0.1813	0.1973	0.1314	1.0000	0.1469	0.5237
Rugby Special	0.3091	0.0637	0.0512	0.2963	0.3412	0.0969	0.3261	0.1469	1.0000	0.1212
24 Hours	0.1242	0.3946	0.2432	0.1403	0.1420	0.2661	0.1221	0.5237	0.1212	1.0000

Factor or component analysis might now be used to clarify the data. Table 14.8 gives the results of a *Principal Component Analysis* as produced by an appropriate computer package. Principal components are linear functions of all the observed variables. They are chosen so that the *first* principal component is the linear function which accounts for the greatest possible part of the total variance in

**TABLE 14.8   Factor Loadings of the First Four Principal
Components**
(Correlations of the components with the original variables)

	Principal Components				
	I	II	III	IV	etc.
Professional Boxing	.7	-.3	.0	.0	.
This Week	.4	.6	-.3	-.3	.
Today	.3	.5	-.4	.7	.
World of Sport	.8	-.3	.1	.1	.
Grandstand	.8	-.3	-.1	.0	.
Line-up	.3	.4	.7	.3	.
Match of the Day	.7	-.4	-.1	.0	.
Panorama	.5	.5	.0	-.4	.
Rugby Special	.5	-.2	.3	.0	.
24 Hours	.5	.6	.0	-.3	.
% Variance accounted for	32%	18%	8.8%	8.5%	.

the data, i.e. correlates most highly with the observed variables. In our case Component I accounts for 32 percent of the total variance of the ten observed variables. From the remaining variation a second principal component is extracted, uncorrelated with the first. It in turn accounts for more of the remaining variation than any other component could, here 18 percent. This continues for the third and fourth components, etc. The possible number of principal components equals the number of observed variables.

In our example there would be ten principal components. The third and later principal components however each account for less than 10 percent of the total variance. They would usually be discarded because, in effect, each accounts for less of the observed correlation than one of the original variables would, in standardized form. (This is often put in terms of the 'Eigenvalue' of a component being less than 1.) In this sense there is a marked reduction in the number of variables: from ten observed variables to two components. But the criteria which factor analysts use for determining the number of components or factors are widely regarded as quite arbitrary.

The two new components now need to be interpreted. This is done simply by inspecting the numbers in Table 14.8, usually referred to as 'factor loadings'. They are the correlations between the new components and the original variables.

The interpretations might be as follows:

Component I is positively correlated (.3 or more) with all ten variables. It might represent how much people like television, since the observed variables represent the numbers of people who said they 'really like to watch' each program.

Component II is correlated positively with This Week, Today, and other current affairs programs (but only slightly higher than Component I). It is correlated negatively with Professional Boxing, World of Sport, and other sports programs, and hence might imply a 'disliking' of these. It therefore appears to differentiate between the two types of programs.

Factor Rotation

Principal components are often not easy to interpret. It is therefore common to derive new components, usually as some linear function of the principal components. Thus a new component might be defined as

$$\text{New component} = p\text{I} + q\text{II} + r\text{III} + \ldots$$

where p, q, r, etc. are new numerical coefficients. This procedure is usually called a 'rotation' (since the technical procedures were initially graphical). It is a major part of modern component and factor analysis. There are many different procedures on the market, often giving different results, so that the procedure again becomes arbitrary.

Table 14.9 shows the results of a *Varimax* rotation—a popular procedure. Here it is applied to the first two Principal components. The aim is to produce new factors each of which has very varied correlations with the original variables—some large and the rest near zero. This can make it easier to name the factors.

TABLE 14.9   **Factor Loadings for a Varimax Rotation**
(Based on the first two principal components only)

| | Varimax Factors | |
	S	C
Professional Boxing	.7	.1
This Week	.1	.7
Today	.0	.5
World of Sport	.8	.1
Grandstand	.8	.1
Line-up	.0	.5
Match of the Day	.8	.0
Panorama	.2	.7
Rugby Special	.5	.1
24 Hours	.1	.8

Factor S in Table 14.9 is highly correlated with just five of the programs. All of these deal with sport. Factor S therefore seems to represent the liking of sports programs. It can be called a 'Sports Factor'. Similarly, Factor C is highly correlated with liking the five current affairs programs. It can therefore be called a 'Current Affairs Factor'. The pattern of the correlations or 'loadings' can be clarified by rearranging the order of the programs, as in Table 14.10. (The results are not always this clear-cut.)

**TABLE 14.10   The Varimax Solution with
the Variables Re-ordered**
(From Table 14.9)

| | Varimax Factors | |
	S	C
World of Sport	.8	.1
Match of the Day	.8	.0
Grandstand	.8	.1
Professional Boxing	.7	.1
Rugby Special	.5	.1
24 Hours	.1	.8
Panorama	.2	.7
This Week	.1	.7
Today	.0	.5
Line-up	.0	.5

Naming the factors—here 'Sports' and 'Current Affairs'—is often the end of the analysis. But sometimes 'factor scores' are calculated. They are values which represent the new factor variables as functions of the original variables (as noted when we introduced factor analysis at the beginning of this section):

$$S = a_s(\text{PrB}) + b_s(\text{ThW}) + c_s(\text{ToD}) + \ldots .$$

$$C = a_c(\text{PrB}) + b_c(\text{ThW}) + c_c(\text{ToD}) + \ldots .$$

Here a_s, b_s, etc. are new numerical coefficients that can be calculated from the factor loadings.

Such factor scores might then be used in further analyses. For example they might be used to differentiate the viewing choices of light and heavy viewers, or of people with and without a university education. One can now do this in terms of just two new variables instead of all ten original observed ones. This typifies the parsimony that is often aimed at by factor analysts.

However, all ten original variables are still needed to calculate the factor scores. Much of the analysis is also arbitrary and the new variables are more complex and less well understood (being new) than the original ones.

Correlational Clusters

As we have seen, it is usual to interpret a factor in terms of any high correlations it has with some of the original variables. This should mean that these particular variables are themselves relatively highly correlated with each other. A 'Sports Factor' would make little sense unless people who liked one sports program also tended to like the others.

Factor analysis should therefore usually amount to finding clusters or groupings in the original correlation matrix. These clusters should show up in the original correlations (Table 14.7) by rearranging the variables according to the factor loadings. This is done in Table 14.11, where the correlations have been rounded to one digit for additional clarity.

TABLE 14.11   The Correlations for Ten TV Programs Rounded and Re-ordered

Programs		WoS	MoD	GrS	PrB	RgS	24H	Pan	ThW	Tod	LnU
World of Sport	ITV	.	.6	.6	.5	.3	.1	.2	.1	.1	.1
Match of the Day	BBC	.6	.	.6	.5	.3	.1	.1	.1	0	0
Grandstand	BBC	.6	.6	.	.5	.3	.1	.2	.1	.1	.1
Prof. Boxing	ITV	.5	.5	.5	.	.3	.1	.2	.1	.1	.1
Rugby Special	BBC	.3	.3	.3	.3	.	.1	.1	.1	.1	.1
24 Hours	BBC	.1	.1	.1	.1	.1	.	.5	.4	.2	.2
Panorama	BBC	.2	.1	.2	.2	.1	.5	.	.4	.2	.2
This Week	ITV	.1	.1	.1	.1	.1	.4	.4	.	.3	.2
Today	ITV	.1	0	.1	.1	.1	.2	.2	.3	.	.2
Line-Up	BBC	.1	0	.1	.1	.1	.2	.2	.2	.2	.

We can now see that people who said they liked one sports program also tended to like the others (relatively high correlations). The same is true for the five current affairs programs. The correlations between the two different types of programs are much lower, though still positive at about .1. (Some people like anything.) We can also see minor trends and sub-patterns.

In general, one should be able to see the results of a factor or component analysis directly in the original correlation matrix, at least after the event. Otherwise the validity of the results has to be questioned. Often all that is needed is a re-arrangement of the correlation matrix right at the beginning. This is not difficult if the analyst has some prior knowledge of the subject-matter.

14.4   Other Multivariate Techniques

Other statistical techniques of multivariate analysis include discriminant analysis, cluster analysis, and multi-dimensional or non-metric scaling. These techniques are less widely used than multiple regression or factor analysis, although cluster analysis and multi-dimensional scaling are currently fashionable.

Where factor analysis seeks to establish sub-groups of variables which appear to measure the same thing (i.e. which are 'highly loaded' in the same factor), *cluster analysis* generally aims to find groups of items (or people) that are relatively similar to each other. An analogy from biological taxonomy might be the classification of plants or animals into species.

Multi-dimensional scaling seeks to establish configurations from sparse information. For example, it might be used to map a series of towns merely from the road distances between them.

Discriminant analysis is an older technique which aims to establish whether two or more sets of objects differ from each other. A typical example from anthropology might concern two sets of skulls found in different locations, A and B. Are they different or do they stem from the same kind of people? For any one measurement (e.g. depth of cranium) the means of the two observed sets may differ, but there may also be a great deal of over-lap between the two sets of readings. Discriminant analysis aims to combine a number of such measurements (depth of cranium, space between the eyes, etc) as a weighted average to discriminate 'best' between the two sets of data. (Discriminant analysis is like regression analysis except that the *y*-variable consists of two or more separate categories—data for locations A and B—instead of a continuous variable.)

The new 'discriminant variable' could then also be used to diagnose new items as possibly belonging to one of the two sets. This kind of procedure seems attractive for medical diagnosis, for instance. But although more than 50 years old, the technique has found few well-established applications.

14.5   Multivariable Laws

Law-like relationships in science often involve more than two variables. For example Boyle's Law, $PV = C$, a constant, describes how the pressure P and volume V of a body of gas vary together if the temperature is constant. But we also know that if the temperature T varies, so does the constant C. This leads to the much more general *three-variable* Gas Law that $PV = RT$, where R is another constant.

The role of such a law-like relationship is to integrate and summarize known patterns which have been observed in a wide range of different sets of data. In contrast, the aim of the statistical methods of multivariate analysis outlined earlier in this chapter is to discover *unknown* patterns in *new* data.

A third variable can also enter into a two-variable relationships in order to account for some of the residual scatter. This is like the problem tackled in multiple regression: First we fit an equation $y = a + b_1 x_1$, and then we add a term $b_2 x_2$ to account, if possible, for some of the residual deviations $\{y - (a + b_1 x_1)\}$. But with law-like relationships our starting-point $y = a + b_1 x_1$ would be an empirical generalization which has already been previously established across many different sets of data.

14.6   Discussion

Statistical techniques of multivariate analysis, like multiple regression and factor analysis, can provide relatively quick results. They can generally be applied to any given set of data. With computers the arithmetic need nowadays only take seconds rather than days or weeks as in the past. Hence the procedures have become more popular.

However, different versions of the techniques tend to give different results. The analyst's choice may not be easy since he is usually dealing with new data which he does not yet understand and the technical issues are relatively complex. The outcome is often somewhat arbitrary. Using complex analysis procedures does not necessarily produce useful or valid results.

The techniques generally apply to the analysis of a single set of data. They do not therefore readily lead to generalizable findings. Few quantitative results of lasting value have so far been reported.

CHAPTER 14   GLOSSARY

Cluster analysis	Techniques used to define groups of items which are relatively similar.
Component	A linear combination or weighted average of the observed variables.
Component analysis	A procedure used when there are many variables but no single one is of particular interest. It aims to reduce the observed variables to a smaller number of new variables which account for most of the variation in the data.
Collinearity	Situation in multiple regression where some of the x-variables are correlated.
Correlational cluster	Grouping of relatively high correlations in a correlation matrix.
Correlation matrix	A square array of the correlation coefficients between all pairs of the given variables.
Discriminant analysis	A technique used to differentiate between two (or more) groups of items by constructing a weighted average of different measurements. Also used to assign new items to the different groups.
Factor	A more complex version of a component, with an 'error' term added.
Factor loadings	Correlations between the factors or components and the original variables.
Factor rotation	A technique to derive new factors as weighted averages of other factors or components (e.g. the principal components).
Factor scores	The values which represent a new factor as a weighted average of the original variables.
Multi-dimensional scaling	Seeks to establish configurations or 'mappings' from sparse information.
Multiple correlation coefficient (R)	The correlation between the observed y-values and the y-values given by the multiple regression equation.
Multiple regression	A technique used to derive a best-fitting equation for new data when there are a number of variables and one ('y') is regarded as being of particular interest.
Principal components	Linear functions of all the observed variables where the first component accounts for the highest possible share of the observed variance, the second component the next highest share, and so on.

Regression coefficient	The slope-coefficient for each x-variable in a multiple regression equation.
R^2	The square of the multiple correlation coefficient; it equals the proportion of the variance of the y-values accounted for by the multiple regression equation.
Test variables	Term used for the observed variables in component or factor analysis.
Step-wise regression	Choosing successive x-variables for a multiple regression equation by the extent to which each reduces the residual scatter of the y-variable.

CHAPTER 14   EXERCISES

14.1 The multiple regression equation $y = 3 + 5x_1 + 10x_2$ has been fitted to some data.

(a) What does the equation say about how y varies as x_1 or x_2 vary?
(b) Does this necessarily hold for the observed data?

14.2 What does the multiple correlation coefficient R represent? How does R relate to the variance of the deviations from the regression-equation? How does that relate to the observed data?

14.3 (a) Describe the criterion by which a second variable, x_p say, would be selected in a 'step-wise' multiple regression analysis giving $y = a + b_1x_1 + b_px_p$.
(b) Some other variable, x_q, is correlated .9 with both y and with x_1. Would x_q be more likely to be selected as the second variable in a step-wise regression, if x_p also has a correlation of .9 with y but only an r of .5 with x_1?

14.4 The following table gives an extract of the factor loadings from an analysis of consumers' attitudes to breakfast cereals:

UK Housewives	Factor		
	I	II	III
Nourishing	.6	.1	.3
Value for Money	.8	-.2	-.1
Tastes Good	.3	.5	-.2
Fun to Eat	.2	.6	-.1
Good for Health	.3	-.3	.8
etc.	.	.	.

(a) How might you label or name the three factors?
(b) State briefly what the factors and the factor loadings represent in such an analysis.
(c) What are factor scores?

14.5 In the original data for Exercise 14.4, the correlation between Nourishing and Value for Money was .2 and that between Nourishing and Tastes Good was .7. Comment on the results of the particular factor analysis reported in Exercise 14.4.

PART FIVE: COMMUNICATING DATA

In the first four parts of this book we have discussed ways of summarizing numerical data. Now we consider ways of *communicating* such data.

For specialists the most succinct way may be through mathematics, but the rest of us use numbers, charts, and words. The next four chapters set out some guidelines for using these well.

- Chapter 15 notes that numbers are much simpler to use when they are rounded to two effective digits.
- Chapter 16 illustrates how to make tables easier to read and understand.
- Chapter 17 explains why simple patterns can be clear on a chart or graph, but not numerical detail.
- Chapter 18 presents some guidelines for written reports and oral presentations.

CHAPTER 15

Rounding

A major problem with numerical data is that we cannot manipulate long numbers in our heads. The way to overcome this problem is to round each number to two effective digits.

The information lost in dropping the other digits is usually trivial. Even advanced statistical analysis does not depend on the third or fourth digits. Yet the advantages gained by rounding are big.

15.1 Rounding to Two Effective Digits

Consider two numbers like 17.9 percent and 35.2 percent. They are shown to three digits, as often happens with percentages. But it is difficult to divide 17.9 into 35.2 in our heads, or even to subtract them mentally. But we can all do the mental arithmetic if the percentages are rounded to two digits: 18 from 35 is 17, and 18 into 35 is about 2.

We need to round numbers because of the limitations of our short-term memories. We cannot remember numbers of three or more digits if our mental processes are interrupted. Doing mental arithmetic is such an interruption: if we subtract 17.9 percent from 35.2 percent, we can hardly remember one number when we look at the other, let alone the answer as it begins to emerge.

The basic rule when presenting numerical information is to round numbers to two effective digits. 'Effective digits' means digits that vary in that set of readings. We round to two instead of one effective digit to avoid undue rounding errors. With numbers like 17.9 and 35.2 we therefore round to 18 and 35.

But suppose we had index numbers like 117.9, 135.2, 128.6, and 144.3, where the base year is 100. The initial 1's are not 'effective' because they do not vary from one reading to another. The first two digits that vary are the tens and the units. Therefore we would round to 118, 135, 129, and 144. These numbers are still fairly easy to manipulate mentally, just because the initial 1's are all the same.

There is little loss in relevant accuracy. With numbers varying from 118 to 137, a conclusion or decision will not be affected by whether the 118 was really 118.3

or 117.9. Had the numbers appeared like 118, 135, etc. originally, we would not have needed to round them. If we changed them to 120, 140, 130, and 140, we would incur substantial rounding errors. That would be over-rounding—only one effective digit.

Variable Rounding

When individual readings in a set of data differ greatly, variable rounding can be used. This means that different readings are rounded to different numbers of digits. For instance, readings like 233.4 and 34.3 would be rounded to 230 and 34—to the nearest ten and the nearest unit respectively. This is really just rounding to two effective digits.

We are not used to variable rounding from school, but it makes sense. It means that rounding does not really affect the comparison between any two numbers. In our example the number 230 is still about 200 more than 34, and almost eight times bigger. The 3.4 unit change caused by rounding 233.4 to 230 makes less than 1 percent difference to a number as high as that. But had we rounded 34.3 to 30, the 4.3 unit change would have a fairly big effect—over 10 percent. That is why we rounded 34.3 to 34.

By using variable rounding we do not clutter up the high numbers with digits which complicate the mental arithmetic, but still keep the rounding errors of the low numbers down to the odd percent.

15.2   Exceptions and Safeguards

Some people fear that valuable information might be lost by rounding. But the rounding usually leads to only minute differences in the calculations—far less for example than when putting the readings on a graph. Very often the third digit reported even lacks accuracy. And as a safeguard, one can always keep the data with an extra digit or two in an appendix, filing-cabinet, or data bank. This provides reassurance.

There are exceptions to the rounding rule—cases where rounding to two effective digits would cause too great a loss in precision. But the exceptions are usually clear-cut and easy to manage. For example, a third digit may need to be kept if the readings vary only slightly in the first digit—like 223, 246, 291, 318. This exception is a minor extension of what is meant by 'effective digits'. A similar problem occurs in our decimal system with numbers like 108 and 93, both of which are close to 100. Rounding literally to two digits would give 110 and 90 and would lead to too much rounding error. Thus if all the readings are close to 100, one keeps a third digit for the ones just over 100.

Other possible exceptions occur when using large multipliers (as with exchange rates) or when computing compound interest. In the latter case rounding-off errors can accumulate greatly, so one needs to work with more exact figures.

When calculating deviations from a statistical equation or from an average, it may also be useful during the analysis to use a third digit as a clerical aid. Then the deviations will sum to zero when checking the arithmetic. Similarly, accountants usually aim to get exact checks and balances when auditing an organization's books so they want to use more digits. But rounding the readings to two effective digits would not affect any *conclusions* that could be drawn in these cases. Even if one chooses to work with many digits, the readings should be rounded before the results are presented to others.

15.3 Discussion

Using too many digits in numerical information hampers communication. This holds true in all areas—from advanced statistical analysis to journalistic reporting.

For example, when the BBC radio reported in 1980 that men's weekly wages in the UK rose to £124.57 while women's rose to £78.63, the use of too many digits prevented people from understanding the news. The reporter would have been better off saying men's weekly wages rose to about £125 while women's rose to almost £80. Then listeners would for example have appreciated more easily that the men were earning about 50 percent more than the women.

Mentally rounding long numbers to two effective digits is the only way most of us can cope with doing arithmetic in our heads. But rounding the written numbers themselves greatly helps the reading of data. It also helps recall when we look back at the numbers. This applies not only when communicating data to others, but also in one's own analyses.

Little is lost by rounding because we seldom need more than two effective digits. We would not draw different conclusions from seeing the figures as £125 and £79 instead of £124.57 and £78.63, nor would anyone actually get paid more. Sophisticated analysis does not hinge on the third or fourth digit in each number, but on understanding a wide range of *other* related information.

In analyzing weekly wage-rates, for example, we would examine trends over the years, in different parts of the country, and in different countries. We would look at wages for different age groups and occupations, try to allow for over-time, and so on. This would involve handling hundreds and possibly thousands of readings, not just the couple given in the BBC announcement. One certainly would not interpret each number individually down to the nth digit.

If one uses more precise numbers—like £124.57 and £78.63—one has to pay the price of their being less readable and less memorable. When people cling to more than two effective digits, it is usually because they do not know what the figures mean or how they are to be used, but at least they are precise.

CHAPTER 15   EXERCISES

15.1   Round the following weekly production figures and comment (from *Business Week*, 24 April, 1971):

	Crude oil*	Automobiles
Year ago	11 014	138 838
Month ago	10 556	190 160
Week ago	11 325	146 400
Latest week	11 132	150 941

*In '000 of bulk barrels

15.2   Assess the maximum rounding errors in Exercise 15.1 as a percentage of the average figure in each column and of the range of variation. Comment.

15.3   Without pencil and paper or calculator, give the difference between 372.6 and 821.4. Roughly what is the ratio of the two figures?
Compare the above tasks with doing mental arithmetic using rounded figures like 370 and 820.

15.4   The following annual figures give consumers' expenditure in the UK (in £ million):

1970	1971	1972	1973	1974	1975
31 272	35 093	39 864	45 085	51 507	56 649

Use mental rounding to see how the percentage increase from 1970 to 1972 compares with that from 1973 to 1975.
Now write down the figures rounded to two effective digits and mentally compare the percentage increases again. Comment.

15.5   Round the following figures to two effective digits

9.7, 11.4, 36.5, 92.7, 104.8, 732.1, 1937.3, 4532.5.

Does the variable rounding affect any comparisons you might make of the different figures?

15.6   For the five figures

12, 25, 37, 42, 51,

the mean is 33.4. Write down the deviations from this and from the rounded mean 33. Summarize both sets of deviations by working out either the mean deviation or the standard deviation. Comment on the advantages or disadvantages of both types of means.

CHAPTER 16

Tables

Tables display numerical information for the reader's use. Yet many people feel inept when faced with tables. This is mainly because they are badly presented.

There are several principles which can be used to make tables easier to follow. These can be applied when one is preparing tables for formal reports, when arranging data for one's own working tables, and when trying to glean information from badly presented tables. All the rules have the same basic purpose: to bring out the major patterns and exceptions in the data.

Two of the principles have already been discussed in previous chapters, rounding and using averages. Four further precepts are:

(i) Rows and columns should be ordered by size.
(ii) Numbers are easier to read downwards than across.
(iii) Table lay-out should make it easy to compare relevant figures.
(iv) A brief verbal summary should be given for every table.

To show how these precepts can be applied, let us look at Table 16.1 on p. 225. It gives unemployment figures from a typical table of official statistics. This table is not easy to read and understand, for several reasons. The rows are arranged alphabetically instead of by numerical size; the figures are unrounded; the vertical grid lines interrupt our eye movements; two types of figures are shown; and there is no verbal summary to tell us what the main patterns are. With a few changes and relatively little work, the data can be made more comprehensible, as in Table 16.2.

16.1 Ordering by Size

Table 16.2 is easier to read than Table 16.1, mainly because the rows have been arranged by their average numerical size. Ordering rows and columns by size helps the reader to see patterns and exceptions in the data.

Now we can see that unemployment tends to be steady over the years, both where the numbers are high and where they are low. We can also see details better: on average the figures are lower in 1973 and high in 1974, though Connecticut in 1974 was still low, and Florida, Georgia, and Arizona exceptionally high.

The row order in this table put the larger numbers at the top. This is done because we find it easier to subtract numbers that way; e.g. as

$$\begin{matrix} 670 \\ 230 \end{matrix} \quad \text{rather than as} \quad \begin{matrix} 230 \\ 670 \end{matrix}$$

(When we order the *columns* of a table, our perception is not affected by whether we start with the lowest or highest figures on the left.)

Table 16.2 has also been improved in other ways. It uses variable rounding to facilitate our mental arithmetic. It gives averages to provide a visual focus and to help us see the trends better. It has better layout: fewer grid-lines, but with occasional gaps between the rows. These gaps break the figures into high, medium, and low categories and also help us read across the rows. Table 16.2 also divides the amount of information being displayed: the unemployment figures as a percentage of the work force will be shown separately in Table 16.3. This allows the reader to absorb one basic point at a time.

An External Criterion

The row order in Table 16.2 was determined by the figures in the table, i.e. the row averages. It turned out that the order of these unemployment figures largely coincided with the order of the states' population sizes. It may seem obvious that larger states tend to have more unemployed, but it was not obvious from Table 16.1.

We could have used the population sizes of the states as an alternative ordering device. This would have meant arranging the rows according to information outside the table, like a recent census. The advantage of using such an external criterion is that it can provide the same fixed order for many different tables. Keeping to a fixed order is essential when different tables are meant to be compared; e.g. different kinds of social and economic measures for the same states. (With a larger table, or with many such tables, it would also be worthwhile giving an alphabetical key to help us locate the figures for a particular state: e.g. that Alabama is seventh, Arizona eleventh, California first, etc.)

Table 16.3 illustrates the use of an external ordering device. It gives the percentages of unemployed from Table 16.1, but adopts the same row order as Table 16.2. This table is also easier to read than before, even though the pattern is different from that in Table 16.2. In fact, we can see at a glance that there is no consistent downward trend here. We have therefore learned that these percentages do not follow the same pattern as the states' population sizes or their absolute numbers of unemployed.

There is no single rule about the best ordering device to use. The choice varies with the kind of information that is being presented and what is already known about it. But no matter what form a table takes, some numerical ordering is generally better than none.

TABLE 16.1   The Number of Unemployed by States: 1971–1974
(The first fifteen states)

	UNEMPLOYED							
	Number (1,000)				As % of Work-Force			
	1971	1972	1973	1974	1971	1972	1973	1974
Alabama	75	62	62	78	5.5	4.5	4.5	5.5
Alaska	12	13	14	15	10.5	10.5	10.8	10.5
Arizona	32	32	34	49	4.7	4.2	4.1	5.6
Arkansas	40	36	34	40	5.4	4.6	4.1	4.8
California	737	652	615	670	8.8	7.6	7.0	7.3
Colorado	37	35	36	43	4.0	3.6	3.4	3.9
Connecticut	116	121	89	88	8.4	8.6	6.3	6.1
Delaware	13	11	12	15	5.7	4.7	4.6	6.1
D.C.	34	44	59	62	2.7	3.3	4.2	4.4
Florida	135	127	132	208	4.9	4.5	4.3	6.3
Georgia	76	83	81	109	3.9	4.1	3.9	5.1
Hawaii	21	25	24	27	6.3	7.3	7.0	7.6
Idaho	19	20	19	22	6.3	6.2	5.6	6.1
Illinois	240	245	203	223	5.1	5.1	4.1	4.5
Indiana	128	103	101	123	5.7	4.5	4.3	5.2

TABLE 16.2   Unemployed: States Ordered by 4-year Averages

	Unemployed ('000)				
	'71	'72	'73	'74	Av.
California	740	650	610	670	670
Illinois	240	250	200	220	230
Florida	130	120	130	210	150
Indiana	130	100	100	120	110
Connecticut	120	120	90	90	105
Georgia	76	83	81	110	88
Alabama	75	62	62	78	69
D.C.	34	44	59	62	50
Colorado	37	35	36	43	38
Arkansas	40	36	34	40	38
Arizona	33	32	34	49	37
Hawaii	21	25	24	27	24
Idaho	19	20	19	22	20
Alaska	12	13	14	15	14
Delaware	13	11	12	15	13
Average*	110	110	100	120	110

* For the 15 States

TABLE 16.3   Unemployed as Percentage of Work Force
(States ordered by number of unemployed, as Table 16.2)

| | Unemployed as % of Work Force | | | | |
	'71	'72	'73	'74	Av.
California	9	8	7	7	8
Illinois	5	5	4	5	5
Florida	5	5	4	6	5
Indiana	6	5	4	5	5
Connecticut	8	9	6	6	7
Georgia	4	4	4	5	4
Alabama	6	5	5	6	6
D.C.	3	3	4	4	3
Colorado	4	4	3	4	4
Arkansas	5	5	4	5	5
Arizona	5	4	4	6	5
Hawaii	6	7	7	8	7
Idaho	6	6	6	6	6
Alaska	11	11	11	11	11
Delaware	6	5	5	6	5
Average*	6	6	5	6	6

* For the 15 States

16.2   Rows versus Columns

A further principle is that it is easier to read down a column of numbers in a table than across a row. This is because the leading digits in each number are then close to each other for direct comparison, with no other digits in between.

To illustrate, Table 16.4 gives the percentage of exam passes for six subjects at five schools. There is much variation between subjects. But reading along the rows we can see that the percentages for the different schools are mostly very similar: about 50 percent for Biology with an exception for School C, about 20 percent for Chemistry, mostly 70 percent or so for English, and so on.

**TABLE 16.4   Percentage of Students Passing Exams
in Different Schools**

| | School | | | | |
| | A | B | C | D | E |
Subject	%	%	%	%	%
Biology	49	54	72	51	48
Chemistry	18	23	21	19	16
English	73	81	51	69	76
History	45	46	53	47	52
Maths	23	18	25	29	21
Physics	34	30	36	28	42

This pattern is visually clearer in Table 16.5, where the rows and columns have been interchanged. Reading down each column the relative constancy of the percentages is more apparent: about 50 percent in the first column, about 70 percent in the second, and so on. The exceptionally high 72 percent for Biology and the low 50 percent for English in School C also stand out more clearly than in Table 16.4.

TABLE 16.5   Rows and Columns Interchanged

		Bio-logy	Chem-istry	Eng-lish	Hist-ory	Math-ema-tics	Phy-sics
School A	%	49	18	73	45	23	34
" B	%	54	23	81	46	18	30
" C	%	72	21	51	53	25	36
" D	%	51	19	69	47	29	28
" E	%	48	16	76	52	21	42
Average	%	50*	19	75*	49	23	34

* Excluding School C

The figures are easier to read in columns because the leading digits in each number are next to each other. The effect is even more marked with less-rounded numbers.

The column averages also provide a useful visual focus. Ordering the columns by size would make the pattern still clearer but is not essential here. It is not necessary to use all of the precepts every time.

16.3   Table Lay-out

The lay-out of a table should be designed to help the reader. The main criterion is to put numbers that have to be compared close together. The two preceding guidelines were based on that criterion. Using too many gridlines as in Table 16.1 also impedes visual comparisons.

Widely-spaced rows are also highly counter-productive, as we can see in Table 16.6. The eye has to move much more than in Table 16.5, making the pattern more difficult to see and remember. *Irregular* spacing of rows and columns tends to be particularly distracting. Sometimes this is caused by varying caption lengths, but can be avoided by using abbreviations and footnotes. In contrast, occasional *regular* gaps can help to guide the eye and emphasize the patterns (as in Tables 16.2 and 16.3). Single spacing with occasional gaps is an important but easy rule to adopt.

Reducing the size of a table also helps. With typed tables a 10 percent reduction linearly is very effective (and easy to do on some modern xerographic duplicators). It distinguishes the table from the text and markedly reduces the amount of eye-movement that is required in scanning a table, which in turn reduces the strain on our short-term memory.

TABLE 16.6   Double Spacing

		Subject					
		Biol.	Chem.	Eng.	Hist.	Maths.	Phys.
School A	%	49	18	73	45	23	34
" B	%	54	23	81	46	18	30
" C	%	72	21	51	53	25	36
" D	%	51	19	69	47	29	28
" E	%	48	16	76	52	21	42

If different sub-sets of figures in a table show different patterns, it often helps to display each sub-set in a separate table (as we did in Tables 16.2 and 16.3). The distracting effect of juxtaposing different patterns is highlighted in Table 16.7. The eye has to jump columns to compare like with like. It would be better for the reader if the alternate columns were put into separate tables. Then each would have a clear visual structure; comparing corresponding figures in the different tables is then usually quite easy.

TABLE 16.7   Unemployed: Absolute Figures and Percentages Juxtaposed
(The first five states)

States	Unemployed							
	'71		'72		'73		'74	
	'000	%	'000	%	'000	%	'000	%
California	740	8.8	650	7.6	610	7.0	670	7.3
Illinois	240	5.1	250	5.1	200	4.1	220	4.5
Florida	130	4.9	120	4.5	130	4.3	210	6.3
Indiana	130	5.7	100	4.5	100	4.3	120	4.2
Connecticut	120	8.4	120	8.6	90	6.3	90	6.1
etc.	.	.	.	.	.	.	.	.

Labels are another important element in table lay-out. They should be clear and brief. Labels should be given even with rough working tables: looking at the data a year later one can then still see what the figures represent, what the units were, the date, etc.

16.4 Verbal Summaries

A brief verbal summary of a table helps to focus the reader's mind. It is better to say 'Table X shows that the figures do not vary much from year to year', than 'The findings are shown in Table X'. The first summary guides the reader. He can also see more of the detail and be more critical. The second summary leaves each reader to do the analysis himself.

Even difficult tables become easier to follow with a verbal summary. Suppose we look again at Table 16.1 (repeated below). Once we are told that the numbers of unemployed were largest in California and Illinois we can see it. We can even go on to see more of the detail unaided (e.g. that Florida had the third highest number of unemployed, with a big jump in 1974, etc.).

TABLE 16.1 (Repeat)   The Number of Unemployed by States:
1971–1974

| | UNEMPLOYED | | | | | | | |
| | Number (1, 000) | | | | As % of Work Force | | | |
	1971	1972	1973	1974	1971	1972	1973	1974
Alabama	75	62	62	78	5.5	4.5	4.5	5.5
Alaska	12	13	14	15	10.5	10.5	10.8	10.5
Arizona	32	32	34	49	4.7	4.2	4.1	5.6
Arkansas	40	36	34	40	5.4	4.6	4.1	4.8
California	737	652	615	670	8.8	7.6	7.0	7.3
Colorado	37	35	36	43	4.0	3.6	3.4	3.9
Connecticut	116	121	89	88	8.4	8.6	6.3	6.1
Delaware	13	11	12	15	5.7	4.7	4.6	6.1
D.C.	34	44	59	62	2.7	3.3	4.2	4.4
Florida	135	127	132	208	4.9	4.5	4.3	6.3
Georgia	76	83	81	109	3.9	4.1	3.9	5.1
Hawaii	21	25	24	27	6.3	7.3	7.0	7.6
Idaho	19	20	19	22	6.3	6.2	5.6	6.1
Illinois	240	245	203	223	5.1	5.1	4.1	4.5
Indiana	128	103	101	123	5.7	4.5	4.3	5.2

16.5 Discussion

When an analyst presents information in a table, the reader should not be expected to do the numerical analysis himself. That is the presenter's job. He must guide the reader and help him to understand the points being made. This can be done by using the guide-lines discussed here: rounding figures, using row and column averages, ordering rows and columns by size, etc.

Even if a reader only wants to use an isolated figure from a table, he must understand the patterns of the data in order to interpret that figure. For example, if a senator from Colorado were writing a speech about unemployment in his

State, he might want to take the 43 000 figure from Table 16.1. But to interpret that figure he would have to judge whether it was high or low compared to the other States—both in absolute and relative terms (e.g. compared to population size). The layout of Table 16.1 made such interpretation difficult. Tables 16.2 and 16.3 made it easier.

In general, only the broad patterns in data matter when drawing conclusions or following an argument. Hence only extracts or summary figures need usually be given in the main text of a report or in an oral presentation, rather than detailed tables. Some details may be useful for purposes of illustration, e.g. to demonstrate what certain averages represent. But most of the numbers can be left to appendices, data banks, or filing cabinets. (Any table which is effectively part of the main text should however always be given close to where it is referred to, and never at the end of the report.)

Graphs and charts can also prove useful in communicating broad summary patterns. These we discuss in the next chapter.

CHAPTER 16   EXERCISES

(*Note:* If possible, type new versions of the tables in these exercises.)

16.1   The following table gives the results of an opinion survey of UK motorists (e.g. 37 percent of the sample said that it 'Gives good mileage' for BP, 61 percent said so for Esso, and so on).

Percentage of Motorists who Give the Stated Attitudinal Response for a Brand

	Brand						Av.
	BP	Esso	Fina	Jet	National	Shell	
Attitude							
Gives Good Mileage	37	61	8	12	23	85	38
Long Engine Life	12	19	4	5	9	19	11
For fast, with-it people	7	10	1	3	12	11	7
Value for Money	24	38	5	40	15	40	27
Large Company	30	68	5	2	5	60	28
Average	22	39	5	12	13	43	22

(a) Arrange the columns in order of average size. Describe the main pattern and the exceptions in the data.
(b) Can you now see the pattern in the original table?
(c) Now rearrange the rows in order of average size. Comment.

16.2   In tables where time (e.g. successive years) is one dimension, some people like to start with the latest results in the top row, or in the first column. Others like time to run forward, with the earliest year at the top or on the left. Comment.

16.3 The following table gives the numbers of Swedish retail establishments in 1972 by size and type. Improve the lay-out of the table.

Retail Establishments by Size and Type

Sweden 1972	Number of Persons Serving				
	0-4	5-19	20-49	50-99	100+
Department Stores	0	5	76	168	143
Food, Bevs. and Tobacco	19,735	3,068	266	24	5
Textiles and Clothing	8,550	1,562	95	13	4
Other Types	15,065	2,215	164	17	4

16.4 Improve the lay-out of the following table of housing trends in the UK. Are there special difficulties in changing rows into columns?

	1958	1971	1977	1978
Stock of dwellings (Millions)	16.2	19.5	20.9	21.1
% owner-occupied	39	50	54	54
% rented from local authorities	25	31	32	32
Average price of dwelling (£)	3,246	5,650	13,712	15,674
New dwellings completed (Thousands)				
- Public	148	168	170	136
- Private	130	196	143	152

16.5 The main principle for the lay-out of tables is that figures to be compared should be close together. List six specific aspects of table lay-out where this principle applies. Why does it seem to work?

CHAPTER 17

Graphs

Graphs and charts can make simple results clear and memorable. They are usually not as good at communicating quantitative details or a complex story-line, as Figure 17.1 illustrates.

"McBelding certainly has a gift for making cold statistics come to life."

FIGURE 17.1 (Drawing by Stan Hunt © 1975 *The New Yorker* Magazine)

17.1 Showing Simple Results

Earlier chapters in this book have given many examples where a graph showed up the shape of a distribution or relationship. Figure 17.2 gives another

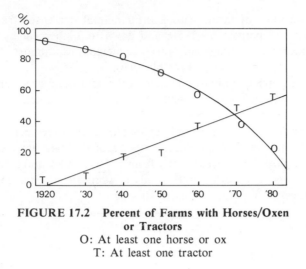

**FIGURE 17.2   Percent of Farms with Horses/Oxen
or Tractors**
O: At least one horse or ox
T: At least one tractor

example—the trends in the incidence of draft animals and tractors in a certain country from 1920 to 1980. The main features of the data are easy to see:

(i) The percentage of farms with at least one draft animal decreased vastly, especially in the later years.

(ii) The percentage of farms with at least one tractor increased steadily over the period.

The message can be summarized in a few words. But the graph makes it more graphic and more memorable, because we can take in and remember shapes and pictures better than words on their own.

17.2   The Numerical Detail

Although Figure 17.2 is effective, the actual *numbers* are not explicit and are therefore not easy to read or to remember. This illustrates the drawback of most graphs and charts: the quantitative detail is difficult to use. For example, in Figure 17.2 how many farms had a draft animal in 1980 compared with 1930?

To communicate the numbers, a well-laid out table will do a better job, such as Table 17.1. Here we can see the numbers, e.g. that

**TABLE 17.1   Percentages of Farms with Draft Animals or Tractors:
1920–1980**

		1920	'30	'40	'50	'60	'70	'80
Farms with at least 1 horse or ox	%	91	89	82	73	56	42	19
at least 1 tractor	%	1	6	20	24	38	52	56

(i) The percentage of farms with a draft animal decreased from about 90 percent to 70 percent by 1950, and down to 19 percent by 1980.

(ii) The percentage of farms with a tractor increased to almost 25 percent by 1950, and to about 55 percent by 1980.

We can also see other quantitative aspects of the data more easily. For instance, there were usually some farms with neither an animal nor a tractor, since only the 1940 column reaches 100 percent.

Even bar charts have difficulty communicating quantitative details. Figure 17.3 illustrates a simple one. The *qualitative* aspects of the data are clear: the total number of unemployed increased greatly from 1966 to 1968, and then only a little after that. This can be expressed more briefly in words, as has just been done. But the graph makes the information more striking.

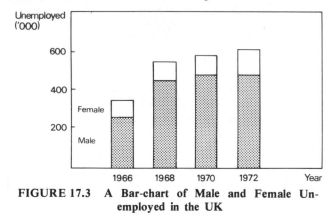

FIGURE 17.3   A   Bar-chart   of   Male   and   Female   Unemployed in the UK

Yet the quantitative details are once again difficult to ferret out. By how much has unemployment risen? How many unemployed were there each year? Did the proportion of *female* unemployed rise less quickly? And so on.

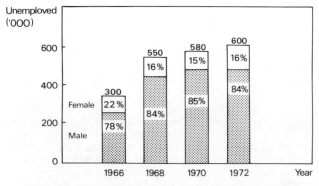

FIGURE 17.4   The Bar-chart with Numbers Added

To help readers see the details, graphs and charts are often embellished with numbers, as in Figure 17.4. But with such a chart one usually looks at the numbers rather than at the sizes of the columns. So it turns out to be little more than a badly laid-out table. Similar difficulties arise with other pictorial methods of data presentation, like pie-charts.

17.3 Complex Story-lines

A graph is effective only if its message can be summarized in a few words. It is wrong to suppose that a graph can simplify *complex* results. As an example, Figure 17.5 shows an illustration adapted from a successful introductory economics text.

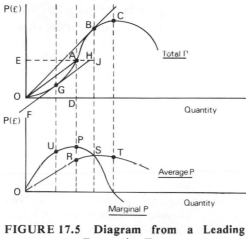

FIGURE 17.5 Diagram from a Leading Economics Text

Even the accompanying words do not help much:

> Average profits are merely total profits divided by the appropriate number of units of output. Thus average profits at point A are equal to OE/OD. But this is merely the slope of the line OA. At any point on a total curve the corresponding average figure is the slope of the straight line from the origin to that point. Given this, we can see. . . .

We may, if we work at it. But a number of simpler graphs would have been more effective, each communicating one idea at a time.

Numerical data for several variables can often be shown better in a well-designed *table*. The variables can be set out distinctly, one per row or column, whereas on a graph they would show up as many over-lapping curves, like in Figure 17.1.

17.4   Discussion

Many graphs are criticized as being misleading. But this is usually true only if the reader does not take enough time or care in looking at them. A much bigger problem with graphs is that many hardly communicate at all. They are often too complex, with no clear visual picture or verbal summary for the reader to take away.

Charts, graphs, and diagrams can be dramatically effective. But they only work well if their story-line is simple and memorable. For this, they must concentrate on broad patterns, shapes, and orders of magnitude—not on quantitative detail.

Graphs are widely misused because they are easy on the eye, so that even bad graphs seem acceptable. We all prefer to look at a graph than at a table. We can follow each line and see where it goes up and then down again. We can also recognize familiar landmarks—like the dip in the 1930s due to the great depression and the blip in 1973 due to the oil crisis. But that does not mean we are absorbing any of the *new* information the graph may be showing.

If one wants to show that something went up or down or stayed the same, and does not need to show by what amounts, then a graph is good to use. But if the amounts are important, then a well-laid out table with rounded numbers will do a better job.

CHAPTER 17   EXERCISES

17.1 Plot a graph showing height against age and weight against age using the data in the following table. What does this graph tell you compared with the table?

Birmingham Boys	Age								
	5	6	7	8	9	10	11	12	13
Av. height h	43	46	48	50	52	54	55	58	59
Av. weight w	42	46	51	57	62	68	73	82	88

17.2 According to the following chart, by how much did imports and exports increase

(a) from 1960 to 1965?
(b) from 1957 to 1970?

Comment.

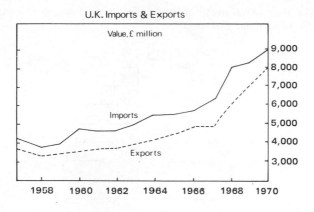

U.K. Imports & Exports

17.3 The following chart gives the age breakdown of the UK population
from 1951 to 1981.

(a) Were there more males than females in total in 1970? If so, how
many more?

(b) How did males and females aged 15–64 compare in 1970?

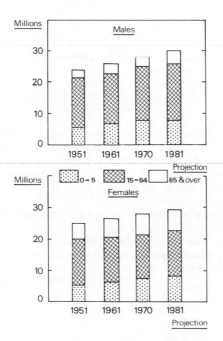

17.4 The chart below gives a more detailed age and sex breakdown of the UK population. Comment.

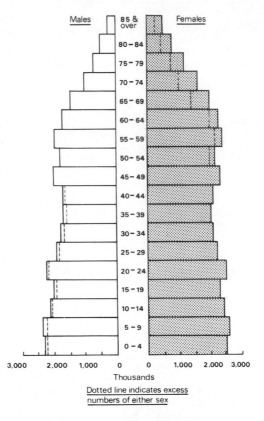

17.5 (a) What characteristics must a graph have to communicate well?

(b) Graphs can be useful to an analyst faced with data that are new to him and also when presenting results to an audience. What do the two situations have in common? What does this tell us about the successful use of graphs?

17.6 Look through a financial newspaper, technical journal, or some volume of official statistics which uses graphs. Classify the first ten graphs you see according to whether each tries to communicate broad qualitative information, quantitative detail, both, or neither, and say whether the message is clear.

CHAPTER 18

Words

We usually explain statistical data by using words, e.g. written reports or oral presentations. Several guidelines can help here. They concern structure, brevity, clarity, the need to revise, and the use of visual aids.

18.1 Starting at the End

The key element of a good report is its structure. This must be geared to what the reader needs.

The traditional way to write a report is to describe chronologically how things happened. A report usually starts with the formulation of the problem and a literature review, and finishes with the results and conclusions. This is logical, but it does not work for the reader. He has to wait until the end for what really interests him: the results and conclusions. These should come near the beginning.

A better structure for a technical report is broadly as follows:

> Title
> Summary
> Introduction
> Main Results and Conclusions

> Detailed Results
> Methods
> Background
> Evaluative Discussion

> Appendices.

The title and summary say what the report is about. But they are too short to inform fully. Next the introduction, also brief, should set the scene, outlining in a page or two the problem, the methods, and the background of the study, but no details. This is followed by the main results and conclusions *in full:* One should never here hold back anything that matters—technical reports are not detective stories. By this time the reader knows what you have found and can stop. (We have all read papers and reports where we seem to have to read on and on before learning the author's purpose or conclusions.)

If the reader is interested, he can continue with the detailed results, the methods in full, and so on. Since he already knows the main results and conclusions, he can now also grasp the detail better.

Good sign-posting is crucial. Clear, succinct headings and sub-headings mainly do this job. A long report also needs a table of contents. With such a structure the reader always knows where he is and what is still to come.

18.2 Brevity

The best way to be brief is to leave things out. In particular, omit what is not essential for the reader. He does not need to know all the steps you took in your work, but only enough to judge the results.

Stating all your main results and conclusions early helps one to be brief. The detailed findings and methods and the exhaustive historical review will no longer seem so necessary. Much can be relegated to appendices or filing cabinets.

Anything which on re-reading seems difficult to express clearly can usually be left out altogether. It probably will not communicate anyway, and since it is not yet clear to you, it is unlikely to be crucial to your main argument.

18.3 Clear Writing

Clarity is next to brevity. Short words, short sentences, and short paragraphs are the key aids here (but with some variation).

The 'Fog factor' is a helpful tool. It is defined as the average number of long words per sentence, 'long' meaning three syllables or more. One can usefully monitor one's writing with samples of about ten sentences at a time.

A fog factor of about 3 is readable. Much academic writing has seven or more long words per sentence and is very heavy. A low fog factor does not guarantee good writing, but a factor of 4 or more provides a warning.

There are two ways to reduce a high fog factor. One is to omit or replace some of the long words (e.g. 'fundamental' by 'deep'). The other is to divide a long sentence into two. But not all sentences should be short. And some words can be quite long—up to an average of about three per sentence!

18.4 More than One Draft

Drastic revision is essential in producing a good report. Professional writers edit and revise as a matter of course and never expect their first drafts to be perfect. So when writing a report we should allow time for revision. (Sometimes there is none, but that does not mean the report could not have been improved.)

One should start with a catalogue of points to be covered. Keeping the report's final structure in mind, as in Section 18.1, these points can be developed into an outline of main headings and sub-headings. This may need several revisions. Thinking through the outline again and again can save much time and effort later on.

We often do not really know how we want to say things, or even what we want to say, until we have written the report out in full as a first draft. Real revision begins here. The draft may be in the wrong order for the reader and need rearranging; often a key finding has crept into a section on details and needs to be put with the main results at the start of the report; we see gaps and discrepancies; we also see better what can be cut.

It is good to leave a draft for some time and then re-read it. ('Did I really write that?') Getting someone else to read it and write 'unclear' or 'too wordy' in the margin also helps us to see the draft in a new light. If the critic queries something, then it is always our fault: it will need to be clarified or cut.

18.5 Oral and Visual Presentations

Oral presentations differ from written reports in two main ways. There is a captive audience, and it must go at the speaker's pace. Since the listener cannot pause or go back to re-read a paragraph, you need to be highly selective and cover only a few points.

Sign-posting is particularly important in oral presentations. Start off by telling what you are going to speak about. As with written reports, give the full account of your main findings and conclusions fairly early on. The listener is fresh then and will have a frame-work for the later details. Aim also to cover all your main points in about two-thirds of your total time. You can then go on with further elaboration, or stop at almost any point if you are running late. Your timing will be perfect.

An audience of less than ten will interrupt and develop a discussion among themselves. That is good, but be prepared for it. An audience of more than thirty will be more like a formal lecture: few if any questions as you go along and possibly not many at the end.

Never read your script, even with a large audience. (Remember other people reading theirs.) Instead, go through your material several times beforehand and prepare a list of key points on a piece or two of paper. Talk to that. (Remember how impressive it is when others do that.) Do not be afraid to pause to look at your notes or to recollect your thoughts. It may seem like ages to you, but it will probably take just a few seconds and the audience will be glad to relax for a moment.

Visual Aids

Aim to use some form of visual aid. It helps the audience to concentrate. Even drawing a rough circle on a blackboard or paper easel and saying, 'This is your market', provides a focus for the listener when his mind has wandered.

Visual aids are good for sign-posting. Early on, show a list of the four or five major points you intend to cover. If possible, leave the list up, or show it again later, as a reminder of the points already covered and those still to come.

To be successful, visual aids require careful preparation and rehearsal. You must always try them out, preferably in the room where you will speak. (There may be no electric socket nearby.) Check whether they are readable from the back of the room.

Flip-charts and transparencies for an overhead projector are flexible: one can write on them and go back and forth fairly easily. But they usually work only for fairly small audiences. Slides are less flexible and more costly, but also more impressive. They are needed for large audiences.

The content of your visual aids must be simple, with few words and fewer numbers. Show results and conclusions, not 'data'. Give the audience time to read what you are showing. Either pause and say nothing, or read the message almost as written. If you explain it in different words, the audience will either listen and not read, or read and not listen.

Hand-outs

Give the audience something to take away. If you give copies of your full report, you can follow a different order of points in your presentation. You can also greatly simplify your explanations ('the full figures are in Section 3.2').

But the hand-out need be no more than your summary points, or copies of any tables and charts you intend to show. (This also provides a fail-safe if the projector does not work.) As listeners we usually like to have the hand-out before the talk, so give it to your audience then.

18.6   Discussion

Most of us are amateurs at communicating. We therefore need to work hard at it.

The experience we had at school was probably quite wrong for report writing. Our school essays were mainly learning devices for our own benefit and they only had to communicate with the teacher, a captive audience of one who already knew it all. In contrast, a technical report is usually for a variety of potential readers, with differing levels of interest and prior knowledge. Some of them will read the report several times; others will stop after the main results and conclusions; some will not read it at all.

Readers need good structure and sign-posting to allow them to choose which parts of the report they want to read. They will read more if you make the report brief and use simple language. The emphasis must not be on what you want to say, but on what the busy reader needs to know.

CHAPTER 18   EXERCISES

18.1   State the advantages and disadvantages of a report which gives all the main findings near the beginning.

18.2   (a) What are the benefits of brevity? What problems are there in achieving it?
(b) Describe three ways of doing so.

18.3   Reduce the following passage by about a third:

'A course on marketing strategy has become almost essential in any business studies curriculum both in the UK and the US. The first real course textbook in this area, Abell and Hammond,[1] went out-of-stock on its first print run within six months. There seems little doubt that we have been witnessing one of the major growth areas in marketing teaching.

Why? Kiechel[2] in a well-written *Fortune* article points out a number of salient facts: the business of strategic consultancy has grown rapidly and many have prospered whilst the tools of such consultancy seem to offer a way for senior executives to cope with the increasing complexity of diverse organisations. In short, many approaches—particularly the now infamous market share/growth matrix[3]—have been well publicised and, rightly, our students and their potential employers demand that they are fully conversant with such approaches. In such circumstances, however, there is a high danger that we, as marketing teachers, fail in our proper task in two respects. First, we can tend to teach . . .'

18.4   It is sometimes argued that jargon is a short-hand which is useful in keeping technical writing brief. Does this contradict the use of the fog-factor?

18.5   Reduce the fog factor of the following passage to about 3.

'Accounting has come to occupy an ever more significant position in the functioning of modern industrial societies. Emerging from the management practices of the estate, the trader and the embryonic corporation (Chatfield, 1977), it has developed into an influential component of modern organizational and social management. Within the organization, be it in the private or the public sector, accounting developments now are seen as being increasingly associated not only with the management of financial resources but also with the creation of particular patterns of organizational visibility (Becker & Neuheuser, 1975), the articulation of forms of management structure and organizational segmentation (Chandler & Daems, 1979) and the reinforcement or indeed creation of particular patterns of power and influence (Bariff & Galbraith, 1978; Heydebrand, 1977). What is accounted for can shape organizational participants' views of what is important, with the categories of dominant economic discourse and organizational functioning that are implicit within the accounting framework helping to create a particular conception of organizational reality. At a broader social level . . .'

Say what, if anything, has been lost.

18.6   When planning an oral presentation of a study on which you are working, what factors do you need to consider if your audience will be
   (a) A small group of colleagues next week?
   (b) A conference in six months' time?

PART SIX:
EMPIRICAL GENERALIZATION

We cannot interpret a numerical finding in isolation but need to see how it compares with other findings. Such comparisons also tell us whether the finding generalizes. Chapter 19 discusses how observable generalizations lead to the lawlike relationships of ordinary science and technology.

We also want to know *why* such results occur, i.e. theory which explains how the findings might arise. But as Chapter 20 shows, establishing underlying mechanisms and causal connections is not easy or certain.

Chapter 21 describes observational studies and experimentation. These are our main forms of data collection.

CHAPTER 19

Description

So far in this book we have concentrated on reducing a given set of data to a simple summary. This enables us then to compare different data sets and to start drawing more general conclusions.

In this chapter we consider the process of comparison and the ensuing results, i.e. the empirical generalizations and law-like relationships of science. These are descriptive results which are known to hold under a wide range of different conditions.

19.1 Compared with What?

The way we view an observed fact depends on how it fits in with other knowledge. If 40 percent of people in an opinion poll say 'The President is doing a good job', is that high, low, or normal? As *many* as 40 percent, or *only* 40 percent?

If last year's figure was 25 percent, the President is now more popular. But last year's 25 percent might have been a freak. To judge this, we need to examine a range of results over a series of years. If they are steady at roughly 25 percent, we know where we stand: 40 percent is high. But if they vary greatly (25 percent last year, 60 percent the year before, 10 percent before that, and so on), we need more understanding of the factors involved. We need studies to assess how public opinion is affected by an election, by the *prospect* of an election, by which party is in power, by the state of the economy, by some foreign-affairs upset, and so on. We need to compare the results of a range of studies before we can really interpret a particular finding like 40 percent.

Developing such comparative analyses from scratch all the time would be intolerable. We cannot collect and analyze hosts of new data to interpret every single figure that we come across. Instead, we rely on past empirical generalizations and norms. The main role of science is to extend and improve such knowledge.

19.2 How Knowledge Builds Up

Knowledge builds up through empirical generalizations: results which have been shown to hold under a wide range of conditions and which continue to hold. The relationship between children's heights and weights in Chapter 13 was an example.

If the same result is found in two different studies, we have a low-level empirical generalization. In contrast, a discrepancy shows that no simple generalization is possible. The basic strategy is, therefore, replication. This does not mean repeating a study under identical conditions, but under *different* conditions. Repeating it under completely identical conditions would be pointless, since the results would have to be identical too. It would be impossible in any case, since at least time or place must differ.

There are always some differences between one study and another. Instead of trying to sweep them under the carpet, empirical generalizations are established by bringing such differences out into the open: the same result holds *despite* such and such differences in the conditions of observation.

The approach is to start with some apparent regularity or pattern. We then repeat the study under rather similar conditions to see relatively quickly and easily whether or not the result was just an isolated freak. If this replication is successful, we then try to repeat the study under conditions as different as possible, but where we think the same result might still arise. Many different factors can here be varied at the same time. If this replication is also successful, the study has given a powerful further generalization for all these factors: the same result holds despite differences in time, place, type of test material, observer, apparatus, etc.

If the extension is unsuccessful, it sets a limit on the generalization. Faced with such a discrepancy, we need to see whether it too will generalize. First we repeat the study under similar conditions to establish that the discrepancy itself was not a freak. Then we back-track by varying only one or two factors at a time, to pin down where and why the difference occurred. For example, time and the particular batch of test material could be varied, but the apparatus and the observer kept the same.

19.3   Law-like Relationships

As one increases the range of conditions under which an empirical generalization is known to hold and also systematizes the conditions under which the result does *not* apply, a low-level empirical generalization evolves into one of the law-like relationships of science. These are our basic means of describing extensive data succinctly. They also provide the building-blocks of theory.

To illustrate, consider Boyle's Law from physics. This says that for a given body of gas, if the volume V in some suitable container is increased, the pressure P of the gas will decrease in such a way that pressure multiplied by volume will remain approximately constant. Written mathematically, $PV \doteq C$ where C is a constant for that body of gas (e.g. that for the same body of gas $V = 10$ and $P = 29$ in one case, and $V = 20$ and $P = 14$ in another, giving approximately the same value for the product PV). This result is called a law because it holds for a wide range of different sets of data, obtained under different conditions. Since the result has held at different times in the past, we can predict that it will hold again in the future, within the range of conditions already covered. Otherwise some new factor must have arisen.

Such law-like relationships have the following properties:
- They are of limited generality, not universal.
- They are approximate rather than exact.
- They are not necessarily derived from theory.
- They are broadly descriptive and not directly causal.

We now discuss these properties, using Boyle's Law as an example.

Laws are General, Not Universal

Boyle's Law $PV \doteq C$ is an empirical generalization. It has been found to hold for different gases, for different amounts of gas, for mixtures of different gases (e.g. air), for different temperatures, for different apparatus, for different observers, for pressure going up and for pressure going down, at different times of day, 300 years ago and now, and so on.

But the law is not universal. It does not hold when the temperature changes, when there is a chemical reaction or a leak in the apparatus, when there is physical absorption or condensation of the gas, or when we tried to prove the law at school.

Establishing the conditions under which such a law does or does not hold is a slow process. Sir Robert Boyle's initial findings applied only to air, under limited experimental conditions. It took 70 years before the law was generally accepted by the scientific community. Many further studies were needed to establish the full range of the law (e.g. to other gases) and to pin down exceptions (e.g. that certain failures are due to the condensation of traces of water vapor).

Laws are Approximate

The laws of science are never 100 percent exact. This does not mean merely that there are errors of measurement when determining specific readings of P and V, but that such laws are always deliberate over-simplifications. Boyle's Law is defined as holding for perfect gases, and these are defined as substances for which Boyle's Law holds. There are no such substances.

In practice, the advanced laws of science generally contain systematic biases. For example, under certain conditions (like high pressure) P is always overstated by the equation, and more so for some gases than for others. This is partly because it is found that other variables should be allowed for, like the mutual attraction of the gas molecules and the volume they occupy. To allow for these, Van der Waal's equation,

$$(P + A/V^2)(V - B) \doteq \text{Constant},$$

provides two correction-factors, A and B, whose numerical values depend on the gas in question. This equation is still only an approximation. But it holds more closely than Boyle's Law, especially at high pressures. Nonetheless, $PV \doteq C$ often remains a close approximation to the facts and we use it whenever possible. The reason is its much greater simplicity: $PV \doteq C$ can be applied without our also having to measure the size of the molecules and the attraction between them for the particular gas being used!

Laws Need Not Be Based on Theory

Scientific laws are not necessarily derived from any deep theory. A successful theoretical explanation often comes *after* a relationship has been empirically established. We usually have to know that something happens (and how generally it happens) before we can explain it. Typically it took 200 years before there was a convincing theoretical explanation of Boyle's Law, in terms of molecular movements, etc.

Before collecting some facts we may have some theory or hypothesis, e.g. that so-and-so might be related to x, y, or z. But this is often merely a conjecture—a way of focusing on what kinds of data to collect and analyze. Fully-fledged and well-established theories are developed much later.

Laws are Not Causal

An equation like $PV \doteq C$ is not causal. It does not say that pressure causes volume, nor that *changes* of pressure cause *changes* of volume. It merely describes approximately what happens if the piston in an apparatus like Figure 19.1 is made to move down from X to Y. Both pressure P and volume V vary in such a way that P times V remains approximately constant. In itself, Boyle's Law is descriptive. This is not to doubt that there is an underlying causal mechanism (like the moving piston), but it does not appear in the equation.

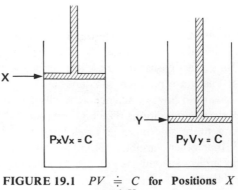

FIGURE 19.1 $PV \doteq C$ **for   Positions** X
and Y

The Problems of Prediction and Extrapolation

A law-like relationship is limited in its generality. It does not necessarily hold outside the range of conditions already tested empirically. Thus the relationship log $w = .02h + .76$ in Chapter 13 held for the heights and weights of a great variety of children of different ages, sex, race, nationality, and so on.

The relationship is restricted to growth under natural conditions. It cannot necessarily be extrapolated to quite different conditions, e.g. to *fast* weight losses or gains due to slimming or overeating. Indeed, we already know from general experience that heights then remain unaltered. Under such conditions the height/weight relationship must therefore be quite different.

Only three races were covered, but this is enough to tell us that race as such does not matter. Thus we may extrapolate or 'predict' that the relationship will also hold for Indian children. If it does not, we know that it must be due either to something specific about Indians (rather than race in general), or to some other special local condition.

19.4   Discussion

Science and technology depend on empirical generalizations, i.e. on results which hold under a wide range of conditions. Such generalizations provide the norms against which to assess new results. They also allow us to predict.

Establishing empirical generalizations is mainly a descriptive function. But it is also the starting-point for understanding. We might have thought that the relationship between children's heights and weights would vary with their sex, race, social class, etc. But since the same law-like relationship log $w = .02h + .76$ holds, we have learned that these factors do not matter. We begin to *understand* the result, i.e. that sex, age, and race do not effect it. The development of theory and explanation is discussed further in the next chapter.

CHAPTER 19   EXERCISES

19.1   Of those who watched the 7 pm News on Channel X last night, some 55 percent had watched the 7 pm News the night before. How would you judge whether this repeat-viewing figure is low or high: *only* 55 percent repeat-viewers, or *as many* as 55 percent?

19.2   A law in physics says that light travels in a straight line. But it does not do so when going through a prism, entering water, or passing near the sun. Does this disprove the law?

19.3   If replication means repeating a measurement under *different* conditions, does it not matter how well the conditions of observation are controlled?

19.4   The sales of different companies are generally highly correlated with their chief executives' pay. Does this mean that increasing a chief executive's salary will increase his company's sales?

19.5   It is common to poke fun at spurious relationships where there is no causal connection, e.g. that life expectancy has doubled since people started to smoke cigarettes. Can you quote a scientific law saying how X and Y are related and where X causes Y, or Y causes X?

CHAPTER 20

Explanation

In the last chapter we discussed how the description of different sets of data can lead to empirical generalizations and the law-like relationships of science. But we are more comfortable with findings we can understand than with ones which simply tell us what happened. We need theory and explanation. We also need to learn more about the underlying mechanisms which probably caused the observed relationships. Such knowledge allows us to predict and to control better.

20.1 Explanation and Theory

Explanations arise by linking separate findings. This greatly simplifies our description of the data. For example it is well-established that when ammonium sulphate is applied to soil, certain plants tend to grow bigger. We explain this finding by saying that the chemical is a *fertilizer*. This simplifies description by linking an isolated finding—that ammonium sulphate makes certain plants grow—to similar findings for *other* fertilizers, broadly defined as materials that can help plants grow.

The explanation, if correct, would also rule out other possible causal mechanisms, e.g. that the chemical in question helps the soil to retain moisture. But the explanation remains superficial. Saying that ammonium sulphate makes plants grow because it is a fertilizer still does not describe how it actually works.

With increasing knowledge our explanations become more detailed. We may be able to say that ammonium sulphate helps things grow because it contains *nitrogen*, which is often deficient in ordinary soil and which plants need. This explanation again simplifies by integrating results for different kinds of nitrogen fertilizers (not just the particular form in ammonium sulphate) and different soil conditions, and by suggesting that the 'sulphate' part of the chemical probably plays no direct role.

The explanation also helps us to understand why fertilizers sometimes do not improve the yield: certain soils may have enough nitrogen. Again, nitrogen applied to potato plants will stimulate a lot of green growth but not bigger potatoes. And so on. But we still do not know how the plants actually absorb nitrogen or what they do with it. The explanations remain descriptive—linking different facts. They still contain unknown mechanisms.

Explanations of increasing range and sophistication tend to be called theories. The advanced theories of science are its greatest achievements. Yet theories are often regarded with suspicion. This contradiction is explained by there being bad theories as well as good ones. The word is often used for hypotheses or models which are not empirically based ('It's only a theory.'). Another cause for mistrust is that even the best theories can never be proven with certainty. They are always over-simplifications and ultimately tend to be superseded.

Scientific Hypotheses

Many theories are too general to test directly. A common procedure then is to deduce a more specific hypothesis and to collect data to test the hypothesis. A general theory might be that cow-manure, unlike artificial fertilizers, does not directly supply nitrogen to plants but provides bacteria which 'fix' the inert nitrogen in the air into a form that plants can absorb. This is difficult to test as it stands.

We can explore the theory by testing the hypothesis that *sterilized* manure would fail to increase yield. Without the speculative theory of how manure might work, one would hardly think of testing it in a sterilized form. The same size increase in yield with sterilized manure as with natural manure would then disprove the 'bacteria' theory. But the alternative outcome (that sterilized cow-manure does not increase yield) would not conclusively prove the 'bacteria' theory. There could be some other mechanism involved. Perhaps the natural manure attracts insects which leads to better pollination or controls pests. We would therefore have to test further hypotheses, e.g. that natural manure is also effective for plants which do not require pollination. Results which are in the expected direction only say that the theory is consistent with the facts.

Good science is a slow process of hopefully piecing together consistent findings and apparently coherent explanations. The normal development of scientific knowledge is one of interaction between fact and theory. Fact collecting is always preceded by some kind of hypothesizing. As noted in Section 19.3, initially this may be only some loose, speculative conjecture that it might be interesting to collect certain facts and not others. When this conjecture turns out to be wrong, it matters little except to the investigator. But as a subject develops, hypotheses increasingly arise from prior knowledge or well-based theory. Failure of such a hypothesis to fit the new facts then implies a discrepancy which needs further investigation.

Statistical Hypotheses

The theory of statistical inference described in Part Three contains many procedures for testing hypotheses. Statistical hypotheses are however narrowly descriptive and differ from those of normal science, which tend to be explanatory or causal.

With a statistical hypothesis our data are based on a sample from a specific population and we merely want to infer what that population would have been like if it had all been measured. The task is to determine, for example, whether or not the use of the fertilizer was related to increased yields for the whole population of crops in a particular area that was sampled. We are not determining *why* it was related.

20.2 Assigning Causes

Explanations and theories imply causal mechanisms. But these are difficult to pin down. For instance, suppose it has been observed that last year school A had more passes per child in a certain maths exam than school B—60 percent against 25 percent. Does this mean that school A is the better school? If so, children would on average do better if they were sent to school A than to school B.

But the observed data—60 percent and 25 percent—do not come from a controlled experiment. School A's pass rate could have exceeded school B's simply because it had more able children in the first place. To try and check on this we need more information.

Table 20.1 shows that the children at the two schools differed in at least one specific respect: their IQs. There was a greater proportion with high IQs at school A. Its higher pass rate could have been due to its having more able children.

TABLE 20.1   The IQs of Children at
Schools A and B

	% of Children at	
	A	B
	%	%
120+	30	10
110-119	40	30
100-109	30	60
- 99	0	0
Average IQ*	115	110

* Using "mid-values" of 125, 115, and 105.

We can allow for the difference in IQ distributions by comparing the pass rates in the two schools for children with the same or similar IQs. Table 20.2 makes the comparisons within certain IQ ranges, from 100 to 109, from 110 to 119, and from 120 upwards.

We see that within each IQ range, school A still had the better pass-rates in the maths exam. But the difference is only about 25 percent on average, slightly more at high than at low IQ levels. It follows that school A's pass-rate was inflated by

TABLE 20.2　　Pass-rates (%) by IQ at Schools A
and B

| | % Passing the Maths Exam | | |
	A	B	A-B
IQ			
120+	86	59	27
110-119	57	31	26
100-109	38	17	21
- 99	0	0	–
Average*	60	25	35 ⟍ 25

*Weighted by % of children in T.20.1

having more high-IQ children, but that there is still something there—a 25- rather than a 35-point difference.

The 25-point difference still need not mean that school A is academically better. It could be that there are *other* prior differences between the children at the two schools. Perhaps those at school A suffer less from absences due to illness, or have been specially coached for the specific exam. So more checks are needed.

If and when we become pretty sure at some stage that school A is academically better than school B, we still need to isolate the particular aspect of school A's performance that appears to be the causal factor. Can we ascribe school A's higher pass-rate to a better teaching method than B has (so that school B and others could also do better with it)? Or to A having better teachers, or a better textbook, or smaller class-sizes, or more maths lessons, or setting more homework, or having supportive parents, or what? The possibilities are myriad.

20.3　A Process of Elimination

To pin down the likely explanation of the different pass-rates, we need to compare many different studies, or separate parts of a given study, like the IQ analysis in Table 20.2. The method is that used in establishing empirical generalizations (Chapter 19), namely seeing whether we can eliminate some of the possible explanations. For example:

- The 25-point difference last year may seem impressive, but if in the preceding or following years school A did no better than school B, there is no general finding to explain. A check whether the basic finding generalizes at all should always come first.
- If school A has higher pass-rates than B in other subjects as well, its success in the maths exam is not simply due to better *maths* teaching, or to coaching for the specific exam. (But A could still just be better at getting its children through their exams, without teaching them anything else.)
- If schools A and B have the same class-size, class-size cannot be a factor.

— If school A's pupils also do more homework than B's in other subjects but do not get better grades, then the higher maths passes cannot simply be due to the different amounts of maths homework.

— If the textbooks of schools A and B were used at *other* schools which showed no differences in *their* pass-rates, the choice between the two textbooks does not appear to matter.

— If children at school A are analyzed by high and low 'parental interest' and both groups get the same 60 percent pass-rate, then having more supportive parents at one school than the other might seem not to be a factor. (But parents at school B might be *very* much less supportive, i.e. altogether outside the range of those of school A: 'What do you always want to have your head in a book for all the time?')

Clear-cut conclusions can be drawn only when there is no difference. For example, if a factor does not vary (like class-size), then it cannot have caused the difference, at least not in any simple way. Or if the results in a comparison agree (e.g. no difference in pass-rates for children with high versus low parental support at school A), then that factor cannot have caused the difference. The process is therefore a *negative* one, of progressively eliminating possible causal factors, rather than of positively proving that X is the cause.

Matching and Standardization

Sometimes short-cuts are used to try to eliminate prior differences between A and B, instead of explicitly examining the individual sub-groups as we did in Table 20.2. This approach is used especially when the sample sizes of sub-groups are small. There are two procedures, matching and standardization.

In *matching*, an equal number of children would be selected in each IQ range from all those available at schools A and B. The pass-rates would then be averaged across all these sub-groups at each school and the two overall averages for A and B compared.

In *standardization*, all the available data would be used, but the results for the children in the various IQ ranges would be weighted so that both schools had the same proportions. The weighted pass-rates at each school would then be averaged and compared. (Although all the observed readings are used, there is still some loss in statistical accuracy because of the weighting.)

It is often claimed or implied that when two groups have been matched or standardized on, say, region, age, and sex, that must make the groups fully comparable. But these matching factors may in fact have had little effect on the variable in question. Only detailed analysis by individual subgroups, as in Table 20.2, will show what the real effects of the matching or standardization factor would be. The procedures are therefore useful only when it is already known from more detailed analyses that the factors in question generally affect the comparison and that allowances must be made for them.

20.4　Critical Scrutiny

In looking for explanations we also need to query whether the data really mean what they say. We must ask how the data were probably obtained. What biases, errors or misconceptions might have been involved? What other checks can one think of?

Suppose a salesman claims that his sales in October went down because the weather turned cold. Is that what usually happens? The explanation can hardly stand if sales the year before went *up* when the weather turned cold, or if *other* salesmen's sales went up this year.

The explanation also depends on how sales are defined. If 'sales' equals ex-factory shipments, they could increase in October because the trade is building up stocks for Christmas, and not because consumers are affected by changes in the weather *now*. Are breakages, faulty items, returns, and pilfering included in sales or not? Variations in these factors can cause trends.

As a second example, suppose a newspaper reports that Dr X has said it is a waste of time to brush our teeth when we get up; that what we have to do is brush them *after* each meal. We must ask ourselves how Dr X could possibly know this. What sort of data could he have collected and analyzed? The number of visits to the dentist over 5 years, using two matched samples brushing their teeth before or after breakfast? It seems unlikely, especially since the article did not even mention it.

Perhaps it just seemed obvious to Dr X that it cannot be very effective to clean one's teeth just before getting them dirty again at breakfast. He therefore said so and the newspaper reported it ('Dr X says . . .'). But are there other possible hypotheses? Perhaps teeth get a deposit of 'plaque', and unless we remove this by brushing our teeth *sometimes*, it retains food particles and sugar?

These examples indicate how we can try to understand the limitations of the statistics we come across. In practice, two things are needed: a critical attitude of mind and some knowledge of the subject-matter, such as how sales are defined or how tooth decay might come about.

20.5　Discussion

The possible mechanisms behind an observed result are usually highly varied. We can try to identify the causes by examining many different possibilities and eliminating those that do *not* correlate with the result in question. There can be no causation without correlation.

This approach usually achieves two things. First, we gain a better descriptive understanding of the observed phenomenon, e.g. that the pass-rates in the maths exam do not differ by the amount of homework done, but do correlate with IQ. Secondly, it should help us to focus on the possible causal factor or factors that might be operating.

But even when all the known factors have been eliminated except for one, say that 'modern maths' was used at school A, there still remains doubt whether some other unknown factor was not really involved. (Perhaps school A has children from larger families than school B, so that more of its pupils can get help from older siblings.)

Isolating causes and developing explanatory theories are slow and difficult processes. There is always doubt. This is a traditional feature of science: we can *disprove* theories but not categorically *prove* them. But the success of science and technology in the last few hundred years has shown that extraordinary progress is possible despite this limitation.

CHAPTER 20 EXERCISES

20.1 What does the following 'explanation' tell us: 'Iron forms rust because it combines with oxygen.'

20.2 The 1980 sales of Product X were expected to go down. Advertising in 1980 was therefore increased. But sales went down nonetheless. The year after, demand for Product X was expected to recover and advertising was reduced. Sales went up. Discuss the causal implications.

20.3 Brand A costs 20 percent more than brand B, but sells twice as much as B. Does this disprove the economic theory that consumer demand for products varies inversely with their price?

20.4 The Research and Development Department in a chemical firm have invented a new germicide which they say is substantially cheaper than its nearest competitors. What questions might you raise about this cost estimate?

CHAPTER 21

Observation and Experimentation

There are two basic ways of collecting empirical data: *observational studies* and *experimentation*. In both cases we can develop empirical generalizations and explanations, although often with great difficulty. Progress usually depends on the successful integration of many different studies. It can often be speeded up by deliberate experimentation.

21.1 Observational Studies

In an observational study data are collected under more or less natural conditions. Earlier examples in this book were data on the heights and weights of children, on people's purchases of corn flakes, and on the incidence of strikes. There is usually little or no direct manipulation of the phenomena, aside from taking the actual measurements. In some cases the data already exist (having for example been collected administratively, like sales invoices in a firm, or income tax returns) and merely need analysis.

The weaknesses of an isolated observational study are immense, especially in assigning causes, as illustrated in the preceding chapter. Progress depends on comparing two or more studies, or different parts of a suitably structured *single* study. This can lead to empirical generalizations, as discussed in Chapter 19, and to the elimination of possible explanatory factors, as discussed in Chapter 20. But the process is usually slow, since it depends on (i) what data are available or collectable, and (ii) our ability to select and hence compare data under different conditions.

Regression towards the Mean

A special danger in observational studies can arise through biases directly caused by how we select items or people to be observed or manipulated. Thus if a production process fluctuates naturally, a manager who intervenes whenever the volume is low will imagine that he is very successful, even though output would have improved anyway.

A common example of selective bias occurs when examining two variables like the heights of fathers and sons. Assuming no general trend in the distribution of heights from one generation to the next, some sons when mature will be taller than their fathers, some shorter, and some about the same. But if we select the tallest fathers, their sons will on average be shorter. And short fathers will tend to have taller sons. This is called 'regression towards the mean' because it looks as though the population is moving or 'regressing' towards becoming more average.

The effect is however spurious: there is no trend in average height or in the scatter of heights. The regression is due to the selection procedure. This can be seen by reversing the selection process: the tallest sons will tend to have shorter fathers, and the shortest sons taller fathers.

Such selection problems can be overcome by forming groupings of fathers and sons based on other variables. These might be different localities, ages at marriage, social or economic classes, ethnic backgrounds, etc., rather than one of the variables to be analyzed (e.g. heights of fathers). The problem is to find variables which effectively differentiate between groups of people of different heights.

21.2 Deliberate Experimentation

Our understanding and control of observed phenomena can be speeded up by experimentation. This means deliberately taking readings under artificial conditions. It helps in three main ways.

First, instead of just waiting for some relevant observations to happen, one may be able to *make* them happen. For example, a physicist who wants to observe a rise in temperature applies heat instead of just waiting for the temperature to rise naturally during the day. This speeds things up.

Secondly, one can vary the test factor to a more extreme extent than normal, like raising the temperature to 200 °C instead of just 20 °C, or more generally, create (and observe) phenomena that would not occur naturally.

Thirdly, one can control some of the other factors in the situation, like the humidity or air currents which might interfere with the main comparisons.

But experimentation still does not provide a direct means of assigning causes. Suppose we want to test a new medical treatment. We deliberately run a clinical trial for 200 patients who are given the new treatment and compare the results with those for a control group of 200 matched patients at the same hospital who are given the traditional treatment. Table 21.1 shows a much higher success rate for the new treatment.

TABLE 21.1   Clinical Trial of Two Treatments

Groups with	No. of Patients	Recovering
New Treatment	200	75%
Trad. Treatment	200	45%

But as in a straight observational study, the results still do not necessarily mean that the new treatment is better. There are many other factors that might have varied between the two groups, causing the different responses.

One possible factor is that doctors generally feel constrained to give each patient the treatment that they judge best for that patient. This might result in the new treatment being given more to patients for whom nothing else has worked, or alternatively more to ones who were expected to do well anyway. (Although promising, the new treatment might have side-effects and one would not want to expose patients to this unnecessarily.) When initial prognosis (i.e. *expected* recovery) is taken into account, the outcomes might look very different, as in Table 21.2.

TABLE 21.2   Results for the Two Treatments by Patients' Initial Prognosis

	No. of Patients	Number who Recovered	Number who Did Not Recover	Percentage who recovered
New Treatment				
Good Prognosis	160	144	16	90%
Poor Prognosis	40	6	34	15%
All patients	200	150	50	75%
Trad. Treatment				
Good Prognosis	80	72	8	90%
Poor Prognosis	160	18	102	15%
All patients	200	90	110	45%

The new treatment was given to more patients with good prognoses (160 out of 200). The actual recovery rate was much higher (90 percent) among those initially rated more likely to recover than among those with a poorer prognosis (15 percent), irrespective of treatment. There is in fact no difference in the results of the two treatments once this factor is allowed for.

Patients in such a clinical trial could (and should) therefore be matched by their initial likelihood to recover. But in practice there could still be many other selection factors that might affect the apparent recovery rates for the two treatments.

These illustrations show how deliberate experimentation is in itself no more foolproof in assigning causes than observational studies. The advantage of experimentation lies in the great freedom it can provide to vary the factors of interest and to control at least some of those that are not.

21.3   Randomized Experiments

A special procedure exists to equalize all possible prior differences between different groups in an experimental situation. This is the *randomized experiment*.

With this method we could take a total of 400 patients, as in the previous trial, but divide them into two groups of 200 at random (e.g. by tossing a coin for each patient). One sub-group would be given the new drug; the other would be used as a control group and given the traditional drug. The possibility of any systematic difference between the two groups—e.g. that one group might be more likely to recover anyway—would therefore literally be left to chance.

With large enough samples there will be only a very small possibility that the two groups differ substantially. With small samples the possibility of bias can be estimated through the statistical tests of significance outlined in Part Three.

The randomized experiment was invented by Sir Ronald Fisher in the 1920s. It has been extended to more than two experimental groups and provides a major short-cut for eliminating prior differences when comparing two or more treatments in an experiment.

A base-line of control readings is however not needed when the situation is already well-understood. For example, if the traditional treatment generally gives a 45 percent recovery rate (at different hospitals, in different years, for different doctors, etc.) we do not need an experimental control group. Unless we are interested in small differences, we can judge the results of the new treatment against this norm. That is why randomized experiments are not always as necessary as one might think.

Factorial Designs

A classical precept in experimentation has long been that only one factor should be varied at a time. But several different factors can be varied in a suitably designed experiment with great gains in efficiency and understanding.

Each level of one factor (e.g. the new or traditional drug) can be applied to two or more sub-groups which differ on *another* factor, e.g. dosage. Instead of simply comparing two groups of 200 patients given the new and traditional drugs, one can randomly divide the total of 400 patients into four sub-groups of 100, as in Table 21.3. Then 100 will get the new drug and a relatively high drug dosage; 100 will get the new drug and a normal dosage; and so on.

TABLE 21.3   The Numbers of Patients in a Two-factor
Design
(100 patients per treatment combination)

| | Drug | | |
	New	Trad.	Total
Dosage			
High	100	100	200
Normal	100	100	200
Total	200	200	400

Such a two-factor design is worth more than two single-factor experiments (i.e. one comparing 200 newly-treated with 200 traditionally-treated patients, and another experiment comparing 200 patients with a high dosage and 200 with a normal dosage). The two-factor design can also tell us whether there is an *interaction* between the two separate factors: treatment method and dosage. Table 21.4 shows how the apparent effect of the new treatment might be bigger at high than at normal dosages—90 percent against 50 percent. Indeed, the response to the traditional treatment did not vary with dosage level (40 percent for each).

TABLE 21.4   Recovery-Rates with an 'Interaction' Effect
(The new drug differs more with a high dosage)

	Drug		
	New	Trad.	All
Dosage			
High	90%	40%	65%
Normal	50%	40%	45%
All	70%	40%	55%

If there had been no such 'interaction' between treatment and dosage, each of the two 'main effects' (new versus traditional drug, and high versus normal dosage) would have been established just as efficiently as in two separate experiments of 200 patients each. The new versus traditional drug comparison in Table 21.4 is still based on two balanced groups of 400 patients, and so is the comparison between high and normal dosage levels. But in addition we would have learned that the new treatment worked equally well for the different dosage levels. We therefore would have more than twice the results for the cost of one of the simpler studies.

When there *is* an interaction, as in Table 21.4, a single new versus traditional comparison at only the normal dosage level would be misleading or at least incomplete. It would also not show that the new treatment is much more effective at a higher dosage.

The use of factorial designs is capable of much elaboration. In particular, the test material (here patients) can be grouped into relatively homogeneous subgroups or strata (e.g. by length and severity of illness, prognosis, etc). When sample sizes are small in the individual 'cells' of such an experimental design, appropriate tests of statistical significance are needed. A procedure known as the Analysis of Variance (touched on in Chapter 10) has been highly developed for this purpose.

Limitations of the Randomized Experiment

Despite its ability to eliminate prior differences in experimental material, the randomized experiment has four major limitations.

First, the observed effect may be due to a factor correlated with the apparent treatment, such as an impurity in the new drug. In particular, the mere use of a new or different treatment might have a beneficial effect. In clinical trials this is sometimes checked by using another random sub-sample of patients who are given an inert 'placebo' to simulate the purely psychological effect of being treated.

Secondly, randomized selection does not eliminate differences between the treatment groups which occur *after* the randomization. Hence 'blind' allocation of different treatments is often used so that participants in the study, including the experimenters, do not know which treatment is given to whom. But this does not guarantee that the different groups get the same handling in all other respects. For example, a new drug may have a side-effect which automatically demands more nursing care, and it could be the extra care which causes the higher recovery rate.

Thirdly, deliberate experimental control, let alone randomization, may be either unethical (e.g. using medical treatment which is not optimal for the individual), or impracticable (e.g. the same teacher using two teaching methods), or altogether impossible (e.g. in most of astronomy). Indeed, most scientific knowledge has had to be obtained without randomized experiments, although these might have led to quicker results.

A fourth limitation of the randomized experiment is that the methodology is constrained to a single experiment at a time. Two or more randomized experiments are no longer a fully controlled experiment.

21.4 A Sequence of Studies

Science and technology advance through integrating the results of many different studies. But a series of studies is an uncontrolled situation. It becomes a large-scale observational investigation. This is so even when each study is itself a fully randomized experiment.

To illustrate, consider three randomized clinical trials of our new drug against the traditional one. The trials are carried out at three different hospitals in the UK, the USA, and France in different years and with different doctors, different patients, different diagnostic methods, different drug suppliers, different nursing levels, and so on. Suppose the three studies gave markedly different responses for the new drug, as shown in Table 21.5.

If the three sets of results had been the same, we would have had a simple and

TABLE 21.5   Recovery-Rates for Three Random Clinical Trials of a New Drug

| | Treatment | | New− |
	New	Trad.	Trad.
UK, 1972	90%	40%	50%
USA, 1976	50%	40%	10%
France, 1981	60%	45%	15%

powerful generalization. But trying to understand why these results differ is a matter of analyzing observational material: three separate experiments which took place more or less independently under three different sets of observational conditions. The big difference is the high apparent response to the new drug in the initial study (90 percent) compared to those in the later ones (50 to 60 percent). Why did that happen? Was it a difference in the selection of patients, a difference in the drug, or what?

Although each study was a controlled randomized experiment, trying to integrate and to understand the different results is not a case of controlled experimentation. The observational approach with all its imperfections remains the bedrock of scientific methodology.

21.5   Discussion

Advanced subjects like physics and chemistry have largely developed through deliberate experimentation. But the results of different studies still had to be integrated through the methodology of the observational approach. This slowly leads to empirical generalizations and law-like relationships, to the elimination of extraneous factors when trying to assign causes, and more generally to explanation, understanding, theory, and control.

Developing appropriate experimental comparisons is often difficult, and sometimes downright impossible. In assessing two medical treatments, we either have to use two different doctors (with all the ensuing lack of direct comparison) or the same doctor. He may be less experienced with one method, and perhaps not enamoured of it. There is no simple way out of this dilemma. But we may still be able to devise certain tests and evaluations and make some progress. Thus if treatment A is used by its most dedicated proponent and B by a doctor who is not greatly experienced with it, and A does less well than B, then we have learned pretty firmly that treatment A is not so good.

Experimental results are usually rather artificial. They can seldom be translated *directly* to real life. There is usually a big gap between laboratory physics and practical engineering, and even between wind-tunnel results and real airplanes. Deliberate proto-type applications and development work under operational conditions are needed to bring scientific research results into technological use. But despite its limitations, the experimental approach has led to an unprecedented growth in our understanding and control of observable phenomena.

CHAPTER 21   EXERCISES

21.1   Eugenics used to gain much support from statistical analyses which showed that parents with high IQs tended to have less bright children. Was this part of a general decline in intelligence?

21.2   A firm has 50 salesmen paid on commission and 200 paid a fixed salary. In 1980, 44 percent of the salesmen on commission exceeded their sales target, whereas only 33 percent of those on salary did so. The data are shown below, broken down also by each salesman's 1979 achievement. Comment on the effectiveness of commission versus salary.

Percentage of Salesmen who in 1980 Exceeded their Sales Targets

	No. of Salesmen	% who Exceeded their 1980 Target
On Commission		
High Sales in 1979	40	50%
Low " " "	10	20%
All	50	44%
Salaried		
High Sales in 1979	20	60%
Low " " "	180	30%
All	200	33%

21.3 Physicists measuring temperatures do not usually have beakers of boiling water and ice at the end of their laboratory bench to check their thermometers. Why is this?

21.4 During a clinical trial of a certain drug, the drug was administered at 'high' or 'low' dosages to two samples of patients. There were also two control groups, one given a dummy 'placebo' and one no treatment at all. Each of the four samples of patients was split into two sub-groups which were given high or normal levels of nursing care. The results below show the percentages recovering (e.g. of those given a high dosage at the normal level of nursing care, 35 percent recovered). Assume that the treatments were allocated to large samples of randomly selected patients. Summarize the results and comment on the effectiveness of the experimental design in leading to firm conclusions.

	Drug Treatment			
	"High" dosage	"Low" dosage	Placebo	None
Nursing Care				
Normal	35%	55%	25%	21%
High	40%	60%	35%	20%

APPENDIX A

Statistical Tables

Tables are given of the Normal, t- and χ^2-distributions, logarithms (for transformation of scale), square roots, and 2000 random numbers of selecting random samples.

Values are given to enough precision for most practical purposes. More detailed figures are available in specialist books of statistical tables or some advanced texts.

Table A1: *The Normal Distribution*
Table A2: *The t-Distribution*
Table A3: *The χ^2-Distribution*

Table A4: *Logarithms to Base 10*
Table A5: *Square Roots*
Table A6: *Random Numbers*

Table A1   The Normal Distribution

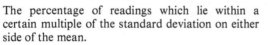

The percentage of readings which lie within a certain multiple of the standard deviation on either side of the mean.

	.0	.1	.2	.3	.4	.5	.6	.7	.8	.9
.0	0.0	8.0	15.8	23.6	31.1	38.3	45.2	51.6	57.6	63.2
1.0	68.3	72.9	77.0	80.6	83.9	86.6	89.2	91.1	92.8	94.3
2.0	95.4	96.4	97.2	97.9	98.4	98.8	99.1	99.3	99.5	99.6
3.0	99.7	99.8	99.9	99.9	99.9	99.95	99.97	99.98	99.99	99.96

Examples:

(a) 68.3 percent of the readings of a Normal distribution with mean μ and standard deviation σ lie between 1.0σ on either side of μ. Just over 34.1 percent lie between μ and $(\mu - 1.0\sigma)$, and just over 34.1 percent lying between μ and $+1.0\sigma$.

(b) About 99.8 percent of the readings lie between $\pm 3.2\sigma$ from the mean and only 1 in 1000 readings lie beyond those limits.

Table A2   The t-distribution

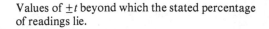

Values of $\pm t$ beyond which the stated percentage
of readings lie.

			Degrees of Freedom												
	1	2	3	4	5	6	7	8	9	10	15	20	30	60	Large
10%	6.3	2.9	2.4	2.1	2.0	1.9	1.9	1.9	1.8	1.8	1.8	1.7	1.7	1.7	1.6
5%	12.7	4.3	3.2	2.8	2.6	2.4	2.4	2.3	2.3	2.2	2.2	2.1	2.1	2.0	2.0
1%	63.7	9.9	5.8	4.6	4.0	3.7	3.5	3.4	3.3	3.2	3.0	2.8	2.7	2.7	2.6
.1%	636.6	31.6	12.2	8.6	6.9	6.0	5.4	5.0	4.8	4.6	4.1	3.9	3.6	3.5	3.3

Examples:

(a) 5% of the readings of a *t*-distribution with 10 degrees of freedom are either less than
−2.2 or greater than +2.2.

(b) For a *t*-distribution with a large number of degrees of freedom (*n* greater than 30),
5% of the readings lie beyond ±2.0.

Table A3   The χ^2-distribution

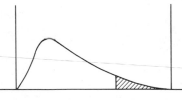

Values of χ^2 above which the stated percentage
of readings lie.

			Degrees of Freedom												
	1	2	3	4	5	6	7	8	9	10	12	15	20	25	30
10%	2.7	4.6	6.3	7.8	9.2	11	12	13	15	16	19	22	28	34	40
5%	3.8	6.0	7.8	9.5	11.1	13	14	16	17	18	21	25	31	38	44
1%	6.6	9.2	11.3	13.3	15.0	17	18	20	22	23	26	31	38	44	51
.1%	10.8	13.8	16.3	18.5	20.5	22	24	26	28	30	32	38	45	53	60

Examples:

(a) 5 percent of the readings of a χ^2-distribution with 6 degrees of freedom are greater
than 13.

(b) .1 percent (1 in 1000) of the readings of this distribution are greater than 22.

Table A4   Logarithms to Base 10

	.0	.1	.2	.3	.4	.5	.6	.7	.8	.9
1.0	.000	.041	.079	.114	.146	.176	.204	.230	.255	.278
2.0	.301	.322	.342	.362	.380	.308	.415	.437	.457	.462
3.0	.477	.491	.505	.518	.532	.544	.556	.568	.580	.591
4.0	.602	.613	.623	.633	.644	.653	.663	.672	.681	.690
5.0	.699	.708	.716	.724	.732	.740	.748	.756	.763	.771
6.0	.778	.795	.792	.799	.806	.813	.820	.826	.832	.854
7.0	.845	.851	.857	.863	.869	.875	.881	.896	.892	.898
8.0	.903	.908	.914	.919	.924	.929	.935	.940	.945	.949
9.0	.954	.959	.964	.969	.973	.978	.982	.987	.991	.996

$$\log 10 = 1$$

Examples:

(a) $\log\ 4 = .602,$

(b) $\log\ 4.6 = .663,$

(c) $\log\ 46 = \log\ (10 \times 4.6) = \log\ 10 + \log\ 4.6 = 1 + .663 = 1.663,$

(d) $\log\ .46 = \log\ (4.6/10) = \log\ 4.6 - \log\ 10 = .663 - 1 = .337.$

Table A5   Square Roots

For numbers from 1.0 to 9.9 and 10.0 to 99.0.

	.0	.1	.2	.3	.4	.5	.6	.7	.8	.9
1	1.00	1.05	1.10	1.14	1.18	1.22	1.26	1.30	1.34	1.38
2	1.41	1.45	1.48	1.52	1.55	1.58	1.61	1.64	1.67	1.70
3	1.73	1.76	1.79	1.81	1.84	1.87	1.90	1.92	1.95	1.97
4	2.00	2.02	2.05	2.07	2.10	2.12	2.14	2.17	2.19	2.21
5	2.24	2.26	2.28	2.30	2.32	2.35	2.36	2.39	2.41	2.53
6	2.45	2.47	2.49	2.51	2.53	2.55	2.57	2.59	2.61	2.63
7	2.65	2.66	2.68	2.70	2.72	2.74	2.76	2.77	2.79	2.81
8	2.83	2.85	2.86	2.88	2.90	2.92	2.93	2.95	2.97	2.98
9	3.00	3.02	3.03	3.05	3.07	3.08	3.10	3.11	3.13	3.15

	0	1	2	3	4	5	6	7	8	9
10	3.16	3.32	3.47	3.60	3.74	3.87	4.00	4.12	4.24	4.36
20	4.47	4.58	4.69	4.80	4.90	5.00	5.10	5.20	5.29	5.39
30	5.48	5.57	5.66	5.74	5.83	5.92	6.00	6.08	6.16	6.25
40	6.32	6.40	6.48	6.56	6.63	6.71	6.78	6.86	6.93	7.00
50	7.07	7.14	7.21	7.28	7.35	7.42	7.48	7.55	7.62	7.68
60	7.75	7.81	7.87	7.94	8.00	8.06	8.12	8.18	8.25	8.31
70	8.37	8.43	8.49	8.54	8.60	8.66	8.72	8.77	8.83	8.89
80	8.94	9.00	9.06	9.11	9.17	9.22	9.27	9.33	9.38	9.43
90	9.49	9.54	9.59	9.64	9.70	9.75	9.80	9.85	9.90	9.95

Examples:

$\sqrt{4}$ is 2.00,
$\sqrt{40}$ is 6.32,
$\sqrt{44}$ is 6.63,
$\sqrt{440} = \sqrt{(44 \times 10)} = \sqrt{44} \times \sqrt{10} = 6.63 \times 3.16 = 20.95$,
$\sqrt{.44} = \sqrt{(44/100)} = 6.63/10 = .663$.

Table A6   Random Numbers

2000 numbers set out in pairs.

```
41 05    41 05    31 87    97 83    98 54    74 53    05 59    17 18    43 12    15 96
14 37    28 51    67 27    89 16    09 71    92 22    23 29    06 37    55 80    03 68
40 64    41 71    70 13    25 96    68 82    20 62    87 17    92 65    46 31    82 88
10 37    57 65    15 62    81 44    33 17    19 05    04 95    48 06    98 69    07 56
57 18    87 91    07 54    11 32    25 49    31 42    36 23    43 86    22 22    20 13

06 12    66 60    93 80    12 23    22 47    47 95    70 17    59 33    43 06    47 43
41 40    24 31    29 85    68 71    20 56    31 15    00 53    25 36    58 12    65 22
95 79    29 19    97 72    08 79    31 88    26 51    30 50    71 01    71 51    77 06
17 79    27 53    85 23    70 91    05 74    60 14    63 77    59 93    81 56    47 34
87 90    68 02    75 74    67 52    68 31    72 79    57 73    72 36    48 73    24 36

30 72    97 57    56 09    07 09    25 23    92 24    62 71    26 07    29 82    76 50
52 23    08 25    21 22    43 31    00 10    81 44    86 38    03 07    53 26    15 87
56 67    16 68    26 95    61 57    00 63    60 06    17 36    37 75    99 64    45 69
35 66    65 94    34 71    31 35    28 37    99 10    77 91    89 41    68 75    18 67
95 91    73 78    66 99    57 04    88 65    26 27    79 59    36 82    53 61    93 78

45 47    35 41    44 22    03 42    30 00    94 03    68 59    78 02    31 80    44 99
35 05    54 54    89 88    43 81    63 61    47 46    06 04    79 56    23 04    84 17
02 82    35 28    62 84    91 95    48 83    47 85    65 60    88 51    99 28    24 39
74 69    00 75    67 65    01 71    65 45    57 61    63 46    53 92    29 86    20 18
08 62    49 76    67 42    24 52    32 45    08 30    09 27    04 66    75 26    66 10

74 01    23 19    55 59    79 09    69 82    66 22    42 40    15 96    74 90    75 89
56 75    42 64    57 13    35 10    50 14    90 96    63 36    74 69    09 63    34 88
49 80    04 99    08 54    83 12    19 98    08 52    82 63    72 92    92 36    50 26
43 58    48 96    47 24    87 85    66 70    00 22    15 01    93 99    59 16    23 77
16 65    37 96    64 60    32 57    13 01    35 74    28 36    36 73    05 88    72 29

76 22    23 87    56 54    84 68    36 60    68 90    70 53    36 82    57 99    15 82
70 72    17 98    70 63    90 32    98 00    82 83    93 51    48 56    54 10    72 32
76 52    26 92    14 95    90 51    12 48    36 83    89 95    60 32    41 06    76 14
88 39    12 85    18 86    16 24    82 04    87 99    01 70    33 56    25 80    53 84
42 66    95 78    58 36    29 98    94 58    16 82    86 39    62 15    86 43    54 31

48 50    26 90    55 65    32 25    87 48    31 44    68 02    37 31    25 29    63 67
96 76    55 46    92 36    31 68    62 30    48 29    63 83    52 23    81 66    40 94
38 92    36 15    50 80    35 78    17 84    23 44    41 24    63 33    99 22    81 28
77 95    88 16    94 25    22 50    55 87    51 07    30 10    70 60    21 86    19 61
17 92    82 80    65 25    58 60    87 71    02 64    18 50    64 65    79 64    81 70

92 47    31 48    75 51    02 17    71 04    33 93    36 60    42 75    07 51    34 87
01 56    63 89    87 43    90 16    91 63    51 72    65 90    44 43    86 59    36 85
99 97    97 78    97 74    20 26    21 10    74 87    88 03    38 33    83 73    52 25
26 54    65 50    98 81    10 60    01 21    57 10    28 75    21 82    08 59    52 18
49 44    29 36    51 26    40 18    52 64    60 79    25 53    29 00    41 27    32 71

66 75    79 89    55 92    37 59    34 31    00 47    37 59    08 56    23 81    22 42
11 26    63 45    45 76    50 59    77 46    86 13    15 37    89 81    38 30    78 68
17 87    23 91    42 45    56 18    01 46    33 84    97 83    59 04    40 20    35 86
62 56    13 03    65 03    40 81    47 54    61 87    04 16    57 07    46 80    86 12
62 79    63 07    79 35    49 77    05 01    43 89    86 59    23 25    07 88    61 29

43 20    45 58    24 45    44 36    92 65    72 63    17 63    14 47    25 20    63 47
34 66    82 69    99 26    74 29    75 16    89 13    29 61    82 07    00 98    64 32
93 13    74 89    24 64    25 75    92 84    03 17    68 86    63 08    01 82    25 46
51 79    80 81    33 61    01 09    77 30    98 08    39 73    49 20    77 54    50 91
30 10    50 81    33 00    99 79    19 70    78 49    19 76    53 91    50 08    07 86
```

APPENDIX B

Answers to Exercises

Chapter 1: Averages

1.1 (a)

	Mean	Median	Mode
(i)	4.1	4	4
(ii)	2.1	1.5	1
(iii)	3.1	4.5	5
(iv)	.1	0	0

(b) (i) 80 percent of the readings are within about 1 unit of the mean of 4.1.

(ii) 80 percent of the readings are virtually at the mean of 2.1 or 1 unit less.

(iii) 80 percent of the readings are either about 2 units above or 2 below the mean of 3.1.

(iv) 80 percent of the readings are within about 1 unit of the mean of .01.

1.2

	Mean	Median	Mode
(i)	74.1	74	74
(ii)	54.0	53	46

For (i), the median is quickest because the readings are ordered by size. (The mode is also quick.) For (ii), the mean is quickest because the readings do not have to be ordered by size.

1.3

	Mean	Median	Mode
(i)	74.0	75	75
(ii)	54.5	55	55

Grouping makes the recording of the data more compact and the analysis quicker, especially for large sets of data. But it can introduce some inaccuracies when calculating summary measures.

1.4 Approximately symmetrical hump-backed data. The three measures are then numerically similar, but the mean is more useful because it is easier to calculate and to use subsequently in other calculations.

1.5 The readings lie within a unit or two of the mean of 3.4, except for the outlier of 54. The 'outlier' looks like a typing error and should probably read 5, 4.

1.6 (a) The weighted average $= (20 \times 30 + 5 \times 80)/(20 + 5) = 40$. The unweighted average is $(30 + 80)/2 = 55$.

(b) No. The mean can be used as a summary measure for different sets of data if the scatter of the readings about the mean has the same shape. The mean can also be a useful *focus* in examining any kind of data, skew or symmetrical. An example is Exercise 1.1(a).

1.7 (a) No. One can only calculate the *mean* of the combined data set. It is $(4 \times 3 + 4 \times 5)/2 = 4$. The median and the mode of the combined set depend on how the individual readings in each data set are distributed.

(b) Even though the medians of A and B are the same as those of P and Q (3 and 5), the medians of the combined data sets differ (3.5 and 4.5). Similarly for the modes.

Chapter 2: Scatter

2.1

	md	variance	sd
Set A	1.0	2.8	1.7
Set B	1.4	3.1	1.8

Set A is easier because the mean was a whole number.

2.2 The two variances are 2.8 and 3.1, as before.
The effect of the short-cut formula is much bigger with the longer numbers in Set B. (Note possible rounding errors.)

2.3 (a) & (b)

	± 1 md	± 1 sd	± 2 md	± 2 sd
Set A:	67%	67%	84%	100%
Set B:	33%	50%	100%	100%

(c) Yes, roughly, even with such small numbers of readings (bearing in mind that the sd is generally somewhat larger than the md).

2.4 (a)

	sds using	
	n	$n - 1$
Set A	1.5	1.7
Set B	1.6	1.8

The differences are roughly 10 percent.

(b) The general effect is $\sqrt{\{(n - 1)/n\}} \times 100$ percent. This is 10 percent, 5 percent, 3 percent, and .005 percent for $n = 5$, 10, 20 and 100.

2.5 (a) Mean $= 5.0$, md $= 1.4$, sd $= 2.2$.
90 percent lie within ± 1 md and also within ± 2 md of the mean.

(b) There are eight readings with a mean of 3.4 and an md of 1.0 (and an sd of 1.3), plus an outlier of 54.

2.6 $$md = 10.4, \quad sd = 13.9.$$

The sd is quicker with a calculator, using the short-cut formula. The md would not be so cumbersome to calculate using the rounded mean of 74 (which still gives an md of 10.4), or if the readings had already been ordered by size (which would make the deviations from the mean simpler to calculate and to check mentally).

2.7 (a) The range is $99 - 44 = 55$. The inter-quartile range is $82 - 66 = 16$. For the range we could pick out the highest and lowest readings with fair certainty from the original data set. For the inter-quartile range the readings had first to be ordered by size.

    (b) Three times the md on either side of the mean is from about 44 to 105. This is similar to the range. Plus or minus 1 md is about 20. This is similar to the inter-quartile range.

Chapter 3: Structured Tables

3.1 (a) Consumption varies from a high of about 64lbs of Potatoes to a low of 13lbs of Fats per head per quarter. QIII is generally about 20 percent below average and QIV 20 percent above.

    (b) The row and column averages show up the above trends. With these in mind, we can then also see that the trends hold broadly for each row or column.

    (c) The deviations (e.g. -2, 2, -11, 13 in the first row) highlight that QI and QII were close to average, QIII low, and QIV high.

3.2 (a) The areas differ systematically, from a quarterly average of 95 for the North to 44 for the South. There is almost no difference between the quarters, except that 1969 QIII in the East and West is about 25 units above average.

    (b) The QIII 'outliers' are not a seasonal effect since they do not occur in the North and South, nor in 1970. Such outliers are best described separately when working out the averages (e.g. 'the East figures are generally about 75, but 100 in QIII of 1969').

    The row and column totals do not provide a useful statistical summary or visual focus because they are in different units. If needed in their own right (e.g. 'total QI sales in 1969 were 271') they can easily be calculated from the averages (e.g. $4 \times 54 = 270$ for QI in 1969).

3.3 Working out the row and column averages as a working tool, we see that they are not typical—there were 'interactions' between countries and time.

	1960	1965	1970	Av.
USA	38,000	49,000	56,000	48,000
Germany	14,400	15,800	16,000	15,400
France	8,300	12,100	14,300	11,800
Italy	8,200	9,000	9,800	9,000
UK	7,100	8,100	7,700	7,600
Average	15,200	18,800	20,800	18,300

Thus the number of reported deaths increased *on average* by a third from 1960 to 1970. But the increase was about 60 percent in the USA and in France, and only about 10 percent elsewhere. The row averages are, therefore, also not typical of the differences between the countries across the decade.

3.4   Writing various averages on the table helps, e.g. those for each pair of plots, for each 'treatment column', and for each fertility level (but for the latter excluding the 'None' column which is clearly much lower). The summary of the main points could be as follows:

  (i) Yields vary with fertility levels (from about 10 to 34 without any fertilizers).
 (ii) At high fertility, all the fertilizer applications increased yield about equally, from 34 to about 48 (i.e. by 40 percent).
(iii) At medium and low fertility levels, the increases were larger

      Medium   from 25 to 43 on average (70 percent).
      Low   from 10 to 30 on average (130 percent).

Nitrogen increased yield by 10 to 20 percent more than did Potash. Applying both increased it even more, especially for the low fertility plots.

3.5   Leaving out the 100s in the diagonal (as systematic outliers), we see that the row averages are generally typical of the individual figures (about 25 percent in the first row, 50 percent in the second, and so on). The figures for viewing the Tuesday Film and News at Ten are exceptionally high.

Chapter 4: Observed Distributions

4.1   (a) 1   2   3   4   5   6   7   8   9

        −   *l* *卅l* *卅* *卅卅* *卅lll* *卅l* *lll* *l*

        − 1   6   5 10   8   6   3   1

     $n = 40$, Mean = 213/40 = 5.325 .

  (b) 0, 0, 0, 0, 2, 4, 6, 9, 13, 15, 17, 18, 20, 29, 33, 43, 46, 51, 54, 61.
     $n = 20$, Mean = 421/20 = 21.05 .

4.2   (a)

1–2	3–4	5–6	7–8	9–10	
1	11	18	9	1	$n = 40$,

0	1–9	10–19	20–39	40–59	60+	
4	4	4	3	4	1	$n = 20$.

  (b) Mean = 54.0 grouped, 53.25 ungrouped; sd 1.69 grouped, 1.67 ungrouped. The loss of precision due to grouping here is small.

4.3   (a) Relative frequencies make it easier to compare the shapes of distributions with different numbers of readings.
  (b) The cumulative distribution for the first data set is:

1–2	–3	–4	–5	–6	–7	–8	–9
Numbers	1	7	12	22	33	36	39	40 ✓
%	2	18	30	55	82	90	98	100 ✓

The second set of readings are so spaced out that they are clumsy to present as a cumulative distribution. But using the grouped distribution in Exercise 4.2(a) gives

0	–9	–19	–39	–59	–79
Numbers	4	8	12	15	19	20 ✓
%	20	40	60	75	95	100 ✓

4.4 (a)

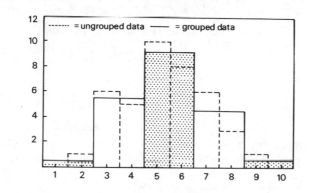

(b) In histograms, the *area* of each block represents the frequencies. For the grouped frequencies, the base of each block is wider so that the height is on average the same as for the individual blocks of the ungrouped histograms.

4.5

$$\frac{7 - 5.325}{1.67} = \frac{1.675}{1.67} \doteqdot 1.0.$$

This says that the value of 7 is about one standard deviation above the mean.

Chapter 5: Theoretical Distributions

5.1 (a) $2\frac{1}{2}\%$; 95%; 5%.
(b) About $.8 \times 4 = 3.2$.
(c) The theoretical distribution extends to plus or minus infinity, but the proportion of negative readings (i.e. more than 5 standard deviations below the mean) would be negligible.

5.2 Set 1 is close to Normal. The distribution is virtually symmetrical, with about equal numbers of readings lying equal distances above or below the mean; 60 percent of the readings lie between ±1 sd from the mean, and 95 percent between ±2 sd.

For Set 2, again 60 percent and 95 percent lie between ± 1 sd or ± 2 sds. But the distribution is somewhat skew, with 50 percent of the readings greater than the mean, 35 percent less than the mean, and 15 percent lying at the mean.

Set 3 is clearly non-Normal. It is reverse-J-shaped, with 30 percent of the readings taking the lowest value.

5.3 The distribution of the widths is skew, with seven out of ten readings lying above the mean of 7.6.

The distribution of *areas* is closer to Normal. Half of the readings are greater than the mean of 70 and half are less. Roughly 68 percent and 95 percent of the readings lie between ± 1 and ± 2 sds.

5.4 (a) The mean 1.6 virtually equals the variance of 1.58. Hence a Poisson could fit.

(b) Using Table 5.6, $e^{-\mu} = e^{-1.6} = e^{-1} \times e^{-.6} = .368 \times .549 = .202$. Therefore

$$p_0 = e^{-\mu} = .202, \qquad p_1 = (\mu p_0)/1 = .323, \qquad p_2 = (\mu p_1)/2 = .258,$$
$$p_3 = (\mu p_2)/3 = .138,$$

and similarly

$$p_4 = .0552, \qquad p_5 = .0177, \qquad p_6 = .005.$$

As percentages (e.g. 20.2 percent for p_0), these figures agree closely with the observed ones.

5.5 (a) The Binomial has a fixed upper limit and could not apply to (i). But it *could* apply to (ii). (There is past evidence that it does!)

(b)

$$\frac{n!}{(n-r)!\, r!} p^r (1-p)^{n-r} = \frac{20!}{18!\, 2!} .1^2\, .9^{18}$$

$$= \frac{20 \times 19 \times \ldots \times 3 \times 2 \times 1}{(18 \times 19 \times \ldots \times 3 \times 2 \times 1)(2 \times 1)} (.01)(.15)$$

$$= \frac{20 \times 19}{2 \times 1} (.0015) = .285, \text{ or 29 percent approx.}$$

(c) The quick way is to use the theoretical Binomial formulae

Mean $= np$ $= 20 \times .1$ $= 2.0$
Variance $= np(1-p) = 20 \times .1 \times .9 = 1.8$.

(These values check with those calculated directly from the bottom line of Table 5.10.)

5.6 The theoretical distribution mostly fits well. There is an excess of people seeing ten episodes and a short-fall of people seeing eleven, but these differences largely balance out, with an observed 8.9 percent and a theoretical 9.0 percent seeing ten *or* eleven episodes. (Also: Do other TV serials follow the same model? Answer: Yes.)

Chapter 6: Probability

6.1 (a) In general not. E.g. people out on Monday evenings would not watch the program. But any adult *randomly* selected would have a .2 probability.
   (b) No. The multiplication rule applies to *independent* events. People who are someone's friends are not independently selected, and may have similar habits or tastes.

6.2 (a) (i) .260; (ii) .240; (iii) 2 × .2499 = .500.
   (b) A family with two children must have either two boys, two girls, or one of each.
   (c) (i) .133 (i.e. .51 × .260); (ii) .118; (iii) 3 × .122 = .366.

6.3 (a) A Binomial distribution with $n = 50$ and $p = .05$, if the incidence of faulty items is independent and highly irregular.
   (b) That faults appear to occur as if at random, and independently of each other.
   (c) A *mixture* of Binomials with different p-values (possibly like the Beta-Binomial distribution).

6.4 The probability of planes arriving is not constant throughout the day—fewer arrive during the night for example. A mixture of Poisson distributions with different mean values for different times of day might work (e.g. something like Negative Binomial Distribution).
   There will also be a minimum interval between successive landings, so that the assumption of independence is not strictly true. But this may have little effect on the data when analyzing the number of arrivals over relatively long time-periods like hours or days.

6.5 The distribution is likely to be skew, with a long positive tail because of occasional major delays. In practice, there would be few if any correspondingly large *savings* in time.

6.6 (a) This should reflect extensive empirical evidence that about 51 percent of newborn babies tend to be boys under a wide range of different conditions.
   (b) Likely to be 'a degree of belief' statement, perhaps roughly influenced by past experience (i.e. that after such statements it *did* rain something like half the time).
   (c) A 'degree of belief' statement, with no direct empirical basis of a sequence of similar events. (But it might be said in the expectation—probably untested—that one would be right 80 percent of the time when claiming a .8 probability for *anything*.)

Chapter 7: Taking a Sample

7.1 (i) Taking every fifth item (reading across in rows) gives 54, 57, 59, 60, 64. This is very biased (the original list being systematically ordered by size). Thus the sample mean is about 59.8 compared with the population mean of 43. The readings are also 'surprisingly similar'. (In practice one often roughly knows the mean and scatter of the population one is sampling.)
   (ii) Using the third row of random numbers in Table 7.1 would identify the following sample readings: 39, 64, 39, 46, 60.
       The seventeenth person (aged 39) has here been picked twice (when sampling

with replacement). The sample, being small, is still not very accurate (a mean of 49, and no one under 39).

(iii) Take one person at random from each column (since the population is arranged in increasing order of age). Using the fifth row of random numbers in Table 7.1 (single digits will do) identifies the following sample: 31, 39, 46, 49, 57. The mean of 44 is closer to the population mean, as it should generally be for a random sample with *effective* stratification.

7.2 Constructing a sampling-frame (i.e. locating and suitably mapping all the trees) could be an impossibly large task. There will also be problems of defining a tree (as against shrubs and saplings) and of classifying trees by type, size, age, etc., for any effective stratification. Some parts of the forest may also be particularly inaccessible (thickets, ravines, etc).

Some kind of *area sampling* (with pragmatic short-cuts) might be feasible. Taking a good sample in such a case would require knowledge both of trees and of sampling.

7.3 There is a low-high-low-high pattern in the data. Systematic sampling (like taking every 200th reading) would therefore give biased results. One also notes that the afternoon readings are higher. But before designing a better sampling scheme we need to analyze data from other weeks to see if these patterns generalize.

7.4 In each case, targets are set for particular kinds of population members. In stratified random sampling the individuals are selected within each stratum at random (from a specified sampling frame or the like). In quota sampling they are selected more subjectively.

7.5 Although different household members tend to differ by age, sex and activity, they all live in the same neighborhood (same shops, same transport system, etc) and they tend to eat the same meals and watch the same television set.

7.6 Drawing a random sample is a very precise procedure, involving explicit sampling frames and the use of random numbers to identify the sample. In contrast, 'drawing a card at random' means that it is drawn *haphazardly*. This does not exclude systematic biases, e.g. that the top and bottom cards would probably not be selected often enough. The true randomness in 'drawing a card at random' should come from the shuffling.

Chapter 8: How Sample Means Vary

8.1 (a) Using sampling with replacement, one sample will consist of two 0s, another of two 6s, etc. The means of the different samples are as follows, with a range from 0 to 6:

Mean:	0	1	2	3	4	5	6	
Frequency:	1	2	3	4	3	2	1	Total 16

This is the distribution of sample means of all possible samples of size 2, also referred to as the sampling distribution of the mean.

Its shape is symmetrical with a peak in the middle. It is too even to be a Normal shape, but is not far from it.

(b) The standard deviation of the sample means is $\sqrt{(40/16)} = \sqrt{2.5} = 1.6$. This is usually called the standard error of the mean. The standard deviation of the population of four readings is $\sqrt{(20/4)} = \sqrt{5} = 2.2$. These results tie in with the theoretical formula $\sigma/\sqrt{n}$ for the standard error, i.e. that $2.2/\sqrt{2} = 1.6$.

8.2 (a) $m = 6$, $s = \sqrt{(90/19)} = 2.2$.
The standard error of the mean $= 2.2/\sqrt{20} = .49$, or .5 rounded.

(b) Although suspiciously symmetrical for a supposedly random sample, the twenty readings look near-Normal. The population sampled is therefore probably near-Normal, and so the sampling distribution of the means for samples of twenty will be virtually Normal. Hence the standard error tells us that 68 percent of all possible sample means will lie within $\pm.5$ of the population mean, and 95 percent within ± 1.

8.3 (a) Substitute the sample standard deviation s for the unknown population σ to give the estimated standard error $s/\sqrt{n}$.

(b) The sampling distribution of the mean has a standard deviation $\sigma/\sqrt{n}$, but σ is unknown. We therefore use the t-ratio $(m - \mu)/(s/\sqrt{n})$ which is known to follow a t-distribution. When n is greater than about 10, the t-distributions are almost identical with the Normal distribution.

(c) 1 percent (from Table 8.2).

8.4 (a) $n = 25$ and $n = 2500$.

(b) Prior knowledge of other similar data, or a pilot study.

8.5 (a) .1

(b) The design factor refers to the ratio of the expected sampling variance of the mean to its variance for simple random sampling. The ratio of the standard errors is therefore $\sqrt{1.9} = 1.4$, so that the estimated standard error is 40 percent higher than for simple random samples, i.e. $1.4 \times .1 = .14$.

Chapter 9: Estimation

9.1 (a) The sample means are 1, 2, 3, 3, 4, 5 with a mean of 3, equal to the population mean $\mu = 3$.

(b) The sample ranges are 2, 4, 6, 2, 4, 2, with a mean of 3.3. This is much smaller than the range of the population itself, which is 6.

These figures reflect two general results: that the mean of a random sample gives the right answer on average (i.e. is an 'unbiased estimator' of the population value), and that the sample range does not.

9.2 (a) The estimated standard error is $4/\sqrt{250} = .25$. The 95 percent confidence limits of the mean are therefore from 24.5 to 25.5.

Statements comparable to 'The population mean will lie between 24.5 and 25.5' will, therefore, be true for 95 percent of all random samples of $n = 250$ from this population.

(b) For 5 percent of the samples the population mean will lie outside their 95 percent confidence limits, but usually only a little outside.

9.3 (a) No. The sample size n has already been taken into account in calculating the standard error and hence the confidence limits.

(b) Neither statement (i) nor (ii) is correct.

Statement (i) is wrong because in the traditional theory of statistical sampling one cannot make probability statements about the possible values of the population mean since this has a fixed (though unknown) value.

Statement (ii) is wrong because the limits 10 and 14 apply only to the sample in question. The observed data for other samples, and hence the comparable calculations, will be numerically different. The proper interpretation of confidence limits is that statements equivalent to saying that the population mean lies between 10 and 14 will be true for 95 percent of all possible random samples.

9.4 (a) The t-ratio is $(25 - \mu)/(4/\sqrt{5}) = (25 - \mu)/1.8$.

(b) The 95 percent t-value for $(n-1) = 4$ degrees of freedom is 2.8. Thus the 95 percent confidence limits for μ are $25 \pm 2.8 \times 1.8$, or from about 20 to 30.

9.5 The finding that g is about 32 feet per second^2 near the surface of the earth is so well established that the new average of 23.4 in London cannot change our belief in the value of 32. One would therefore conclude that something must have 'gone wrong' for the new data, e.g. some radical difference in the conditions of observation, some consistent bias or error in the measurement procedures, or Twyman's Law (should it read 32.4?). Check the typing and if still necessary, collect some more readings in London.

9.6 The table is based on twenty separate random samples. This tells us something about the likely errors due to sampling, i.e. how much different samples can vary. Thus the readings in the body of the table differ mostly by 1 or 2 units from the row or column averages, or from the overall average of 8. At least a part of these differences will be due to sampling errors. The latter will therefore generally not exceed 1 or 2 units.

Chapter 10: Tests of Significance

10.1 The standard error of the sample mean 12 is about .4. Hence m differs by about $2\frac{1}{2}$ standard errors from the usual value of 11 and is significantly different at about the 1 percent probability level.

This says that by the chance errors of sample selection, only 1 in 100 random samples of 220 loans would have means differing by one day or more from the usual loan average of 11 days. It is unlikely that our particular sample was that 1 in 100 sample. It is more likely that the average lengths of loans for *all* books in March was longer than 11 days. Thus we reject the null hypothesis that $\mu = 11$.

10.2 We know already that extra water generally makes plants grow more (at least under broadly normal conditions). There is probably no point in testing the hypothesis that watering has *no* effect.

We can however set confidence limits for the result, i.e. 2 ± 2.4 at the 95 percent level, or from $-.4$ to 4.4. These limits include zero, so it is possible that extra watering here had no real effect. But the best estimate of the effect for the population sampled is a 2 inch increase.

The *practical* significance of such an increase depends on whether the extra 2 inches or so would be worthwhile in terms of the various costs of supplying the extra water.

10.3 Fewer sample values would reach the 1 percent than the 5 percent significance level if the treatment had no special effect for the population as a whole. Thus the sample result would make the doctor less likely to claim such an effect wrongly, i.e. to make a Type I error.

Reducing the chance of making a Type I error increases the chance of making a Type II error. This is to conclude from the sample data that the treatment had no effect even though it had one for the population as a whole.

10.4 The standard error of the mean is $2.2/\sqrt{4} = 1.1$. To assess the significance of the difference $(m - \mu) = 3.6$ for a small sample we have to use the t-distribution with $(n - 1) = 3$ degrees of freedom. The t-ratio $(m - \mu)/(s/\sqrt{n}) = 3.6/1.1 = 3.3$. This is virtually equal to the 95 percent value for the t-distribution and hence is significant.

The only assumption that is being made is that the readings in the population are Normally distributed (or near Normal). This is impossible to assess from such a small sample, but there might be information from other similar data. (For highly non-Normal data the t-distribution would not give the correct probability levels.)

10.5 The null hypothesis for a χ^2-test is that the two factors are not associated, i.e. that the proportion divorced is the same among those with or without college education, other than for sampling errors. (Or equivalently, that the proportion with college education is the same among those who are divorced or not.)

The overall proportion divorced is $45/140 = .32$ or 32 percent. On the null hypothesis, the number of divorced among the fifty-two with college education is therefore about $.32 \times 52 = 17$. It follows that the number *not* divorced amongst the fifty-two with college education is $52 - 17 - 35$.

Similarly, the number divorced with *no* college education is $45 - 17 = 28$, and the number not divorced among men with no college education is $88 - 28 = 60$ (or equivalently $95 - 35$, which is again 60.)

These calculations show how there is only one 'degree of freedom' in the data. Having calculated the number of divorced among those with college education expected under the null hypothesis, the other three numbers in the table follow automatically.

The value of the χ^2 test-statistic

$$\text{Sum} \left\{ \frac{(\text{Observed minus Expected frequency})^2}{\text{Expected frequency}} \right\}$$

is therefore

$$\frac{8^2}{17} + \frac{8^2}{35} + \frac{8^2}{28} + \frac{8^2}{60} = 14.1.$$

This is greater than the 1 percent probability value of the χ^2-distribution with 1 degree of freedom (6.6, see Table 10.8). Hence the observed difference in the proportions of divorced men among those with and without college education (48 percent versus 23 percent) is highly significant and almost certainly not just due to sampling errors.

Chapter 11: Correlation

11.1 (a) $(x - \bar{x})$ $-4, -2,\ \ 1, 2, 3.$ Sum $= 0.$
            $(y - \bar{y})$ $-3,\ \ 1, -2, 3, 1.$ Sum $= 0.$
            $(x - \bar{x})(y - \bar{y})$ $12, -2, -2, 6, 3.$ Sum $= 17.$
        (b) Cov $(x, y) = 17/4 = 4.25.$ var $(x) = (16 + 4 + 1 + 4 + 9)/4 = 8.5.$
            var $(y) = 24/4 = 6.$ $r = 4.25/\sqrt{(8.5 \times 6)} = .60.$

11.2 $r = (137 - 5 \times 4 \times 6)/\sqrt{(114 - 5 \times 4^2)}\sqrt{(204 - 5 \times 6^2)} = .60.$

11.3 s_x is ten times as big as before, s_y stays the same, and cov (x, y) is ten times as big. The correlation coefficient therefore stays the same. (Direct calculation also gives $r = .60.$)

11.4 Both correlations are .8. But in the second study, P varies twice as fast with Q as in the first study. Thus when the correlation coefficients are the same in two studies, it still does not tell us whether or not the two relationships are the same.

11.5 (a) We cannot tell.
        (b) r would be lower if the scatter about the relationship in the UK were larger than in the USA, or if the readings in the UK were more bunched together than in the USA.

Chapter 12: Regression

12.1 (a) $y = 12 + .67x.$
        (b) 6.

12.2 (a) $b = (-20 + 0 + 0 - 6 - 8)/(16 + 1 + 0 + 1 + 16) = -1.$
            $a = 8 - (-1 \times 7) = 15.$ Hence the regression is $y = 15 - x.$
        (b) $b = (246 - 5 \times 7 \times 8)/(279 - 5 \times 7^2) = -34/34 = -1,$ as in (a).

12.3 (a) The deviations $y - (15 - x)$ are: 1, -1,   3, -5,   2 (summing to zero). The variance of the residuals, using $(n - 1)$ as divisor, is $40/4 = 10.$ The rsd is 3.2.
            The two theoretical formulae for the residual variance also give 10:
            var $y - b^2$var $x = (74 - 1 \times 34)/4 = 10.$
            $s_y^2(1 - r^2) = 74(1 - .46)/4 = 10,$ since $r = -34/\sqrt{(34 \times 74)} = .68.$
            (If r has not already been worked out, using the $s_y^2(1 - r^2)$ formula is more lengthy.)
        (b) By plotting the data on a graph or by inspecting the residuals numerically.

12.4 (a) The slope of the regression of x on y is cov $(x, y)/$var $y = -34/74 = -.46$ or about $-.5.$
            The intercept-coefficient is given by $\bar{x} - (-.5\bar{y}) = 7 + .5 \times 8 = 11,$ so that the regression is approximately $x = 11 - .5y,$ which is equivalent to $y = 22 - 2x.$ This differs markedly from the regression of y on x, which is $y = 15 - x.$ One slope is about twice the other.

Chapter 13: Many Sets of Data

13.1 (i) $y = 1 + 12x.$
        (ii) $y = 3 + 11x.$

The results of method (ii) depend less on the means of a single set of readings. It can also be used when the data set with the lowest $\bar{x}$ mean does not have the lowest $\bar{y}$ mean.

13.2 (a) The values of $\bar{y}$ given by the two equations are:

	A	B	C	D	Av.
Observed $\bar{x}$	4	3	1	0	2.0
Observed $\bar{y}$	53	30	12	5	25.0
$1 + 12\bar{x}$	49	37	13	1	25.0
$3 + 11\bar{x}$	47	36	14	3	25.0

The two equations give much the same results (within the range covered by the data). They agree closely with the observed $\bar{y}$ values, the mean deviation being 4.0 in each case. (This is small compared with the fifty point range of the observed $\bar{y}$s, from 53 to 5.)

(b) The deviations of $\bar{y} = 110$, $\bar{x} = 10$ from the two equations are $110 - (1 + 12 \times 10) = -11$ and $110 - (3 + 11 \times 10) = -3$. The second equation, $y = 3 + 11x$, therefore fits both the previous and the new data to within the same limits (a mean deviation of about 4), whereas the first equation, $y = 1 + 12x$, only holds for the initial data. The equation $y = 3 + 11x$ is therefore more general and useful.

13.3 Rather than fit a new equation, one can use prior knowledge. The table below shows how the previous equation of Section 13.4, $\log w = .02h + .76$, holds again, to within the usual limits of $\pm.01$.

	6	8	Age 10	12	14	16	Av.
$\bar{h}$	45.7	49.6	53.5	57.5	62.6	65.7	55.7
$\log \bar{w}$	1.66	1.75	1.84	1.91	2.01	2.08	1.87
$.02\bar{h} + .76$	1.67	1.75	1.83	1.91	2.01	2.07	1.87
$\log \bar{w} - .02\bar{h} + .76$	-.01	.00	.01	.00	.00	.01	.00

13.4 In BGA the emphasis is on seeing whether there is an equation which holds for many different sets of data. This turns on whether the equation holds for the means of the different data sets, at least to a close degree of approximation.

In regression analysis an equation is selected which gives a 'best fit' to a single set of data.

13.5 (a) The discrepancy of the sample means from the equation is $35 - (15 + 10 \times 3)$ $= -10$.

Following the calculations in Section 13.6, the individual residuals $y - (15 + 10x)$ are

$$35, \ 5, \ -5, \ -15, \ -25, \ -55. \quad \textit{Average} \ -10.$$

The deviations from the average of −10 are

$$45, 15, 5, -5, -15, -45. \qquad Average \ 0.$$

Their standard deviation (using the $(n-1)$ divisor) is $\sqrt{(4550/5)} = 30$. The standard error of the average discrepancy of −10 is therefore $30/\sqrt{6} = 12$. Hence the deviation of −10 is not statistically significant.

(b) The correlation of the individual readings in the sample is negative (high y go with low x). This differs from the positive correlation between different data sets which is implied by the slope-coefficient $+10$ in the equation $y = 15 + 10x$.

This also explains why in (a) we saw that many of the individual readings differ so much from the equation (e.g. by up to ± 45 units). Plotting the data with the equation $y = 15 + 10x$ clarifies the picture.

Chapter 14: Many Variables

14.1 (a) Mathematically the equation says that y should vary by 5 units for every unit change in x_1 when x_2 is constant, and by 10 units for every unit change in x_2 when x_1 is constant. A unit increase in both x_1 and x_2 should produce a $5 + 10 = 15$ unit increase in y.

(b) The above need not hold for the observed data. In particular, the observed variation in y for a unit change in x_1 could be higher when x_2 is high than when x_2 is low. A detailed examination of the residuals from the fitted equation is needed to check this.

14.2 The multiple correlation-coefficient R is the simple correlation between the observed y-values and the y-values given by the multiple regression equation for the different observed x-values.

The variance of the residuals from the multiple regression equation is related to R by the equation

$$\text{Variance of residuals} = s_y^2(1 - R^2).$$

This is often said to show the amount of the original variance that is left 'unaccounted for'.

But standard deviations give a more direct measure of the observed scatter. The formula then reads $\text{rsd} = s_y\sqrt{(1 - R^2)}$. For an R of .9, $\sqrt{(1 - R^2)} = .19$ or 44% and so the rsd is still as large as 44% of s_y.

14.3 (a) The second variable x_p is selected from all the measured x-variables (other than x_1) so that the resultant multiple regression equation has the minimum residual scatter (or the highest R) compared with all possible alternatives.

(b) x_p will be selected rather than x_q as the second variable, because x_q is highly correlated with x_1 and x_1 has already been included in the regression equation.

14.4 (a) The three factors can be said to relate to the breakfast cereal being

    I   A good buy ('Nourishing', 'Value for money'),
    II   Good eating value ('Tastes good', 'Fun to eat'),
    III Good for health ('Good for health').

(b) The factors are new variables. The factor-loadings are their correlations with the initial variables, which is all we know about the factors at this stage of the analysis.

(c) The factors are defined as weighted averages of the initial variables. Calculating these gives a 'factor score' for each individual in the study. The factor scores could then be analyzed further to tell us more about the factors (e.g. how the scores relate to the ages of the consumers, the sizes of their families, what they buy, etc.).

14.5 There are many kinds of factor analyses available. They can give different results.

It looks as though the analysis reported in Exercise 14.4 is peculiar. Thus 'Nourishing' and 'Value for money' are only correlated .2 and yet have been shown up as representing the same factor 'I'. In contrast, 'Nourishing' and 'Tastes good' are relatively highly correlated to the extent of .7, and yet appear as major items in two *different* factors. This illustrates the need to check the results of such relatively complex analyses back against the original data.

Chapter 15: Rounding

15.1

	Crude Oil (Million)	Autos ('000)
Year Ago	11.0	139
Month Ago	10.6	190
Week Ago	11.3	146
Latest Week	11.1	151

Both sets of figures are shown to three digits (not two) because the initial digits do not vary. Even so, the oil figures are rounded to only one *effective* digit, but the third digit seems enough to show the small variations which occurred. (In practice one would check with other weeks and years to see if this small variation is general. If there is more variation the apparent over-rounding here would be fully justified.)

15.2 For oil, the maximum rounding error is .04 million bulk barrels for the 'a month ago' figure. This is less than $\frac{1}{2}$ percent of the average weekly production of 11 million. But it is as much as 6 percent of the very limited weekly variation here (which is .7 million in the rounded figures). For autos the rounding errors are about $\frac{1}{2}$ percent or less either way.

15.3 The difference is about 450 and the ratio is just over 2, or 2.2 more accurately. With the rounded figures 370 and 820, typical arithmetic might be: twice 370 is 740, which leaves 80. This is about twice 37 (i.e. 370 divided by 10). Hence the answer is about 2.2. The mental process is easier using figures which are already rounded on paper.

15.4 Rounding mentally to two digits, we have to work out in our heads that the 1970 to 1972 increase is from 31 to 40 thousand, which is 9 over 31, a bit over 25 percent but less than a third (25 percent of 31 would be about 8, a third about 10). The 1973 to 1975 increase is from 45 to about 57 thousand, which is 12 over 45 or also a bit less than a third (15 out of 45 would be a third).

The rounded figures are:

1970	1971	1972	1973	1974	1975
31 000	35 000	40 000	45 000	52 000	57 000

The mental arithmetic is now easier (probably because one can glance back at the numbers and remind oneself what they were in *rounded* form).

For more precise comparisons (if needed) one probably has to use a calculator—the two increases for the rounded figures are 29 percent and 27 percent. (The rounding itself has affected each percentage increase by only about 1 or 2 points.) For many purposes one needs only to know that such increases were roughly the same (e.g. look linear on a graph); otherwise a third digit would have to be used.

15.5   9.7, 11.4, 37, 93, 100, 730, 1900, 4500.

The rounding does not affect the differences or ratios between pairs of figures that are very different. But by using *variable* rounding even the comparisons of figures close together are hardly affected. (Whether to round the 11.4 to 11 depends on the context. When two numbers are close to 10, like 9.7 and 11.4, one may keep an extra digit.)

15.6   The deviations from the means are:

$$m = 33.4: \quad -21.4, -8.4, 3.6, 8.6, 17.6, \quad \textit{Average} \quad 0.0$$
$$m = 33: \quad -21, \quad -8, \quad 4, \quad 9, \quad 18. \quad \textit{Average} \quad .4$$

The rounded deviations do not average exactly to 0.0, which is a nuisance for a precise check on one's arithmetic. But working out the deviations to one decimal place is more tedious, and the extra digit does not affect one's interpretation of the results, e.g. that the deviations range from about −21 to +18.

Chapter 16: Tables

16.1   (a) The table with the columns in order of average size (and with the rows also ordered by size—see (c) below) is:

Attitude	Brand						Av.
	Shell	Esso	BP	Nat	Jet	Fina	
Gives Good Mileage	85	61	37	23	12	8	38
Large Company	60	68	30	5	2	5	28
Value for Money	40	38	24	15	40	5	27
Long Engine Life	19	19	12	9	5	4	11
For fast, with-it people	11	10	7	12	3	1	7
Average	43	39	22	13	12	5	22

This shows that the attitudinal responses tend to decrease together from Shell to Fina, with three exceptions: Esso and Shell for 'Large company', Jet for 'Value for money', and National for 'Fast with-it people'. (Knowledge of the UK gasoline market at the time would show that Jet was a cut-price brand and that National had used commercials showing young couples driving in sports cars.)

(b) The main pattern can now also be seen in the original table (e.g. that the figures for Shell and Esso are high), but it is less easy and the exceptions are not so apparent.

(c) Rearranging the rows in order of average size makes the table still easier to follow, although the effect is not as dramatic as for the columns.

16.2 The way that time is made to run in a table does not greatly affect our perception of the data. Different people or organizations have different habits, and it is better to keep to whatever format is familiar. But for the *rows* of a table it is best to keep the larger numbers at the top.

16.3 Arranging the rows by average size and using variable rounding gives

Sweden 1972	Number of Persons Serving					Total
	0-4	5-19	20-49	50-99	100+	(Rounded)
Food, Beverages and Tobacco	20,000	3,100	270	24	5	23,000
Other Types	15,000	2,200	160	17	4	17,000
Textiles and Clothing	8,600	1,600	95	13	4	10,000
Department Stores	0	5	76	170	140	390
Total (rounded)	44,000	6,900	600	220	150	51,000

This lay-out makes the pattern in the data stand out more clearly, although the position of 'Other Types' may strike one as odd. (A footnote saying 'Other than Department, Food, and Textile stores' would overcome this.)
Changing rows into columns also has a very graphic effect·

Sweden 1972	Dept. Stores	Food, Bev. & Tob.	Other Types	Text. & Clothing	Total (Rounded)
Number Serving					
0- 4	0	20,000	15,000	8,600	44,000
5-19	5	3,100	2,200	1,600	6,900
20-49	76	270	160	95	600
50-99	170	24	17	13	220
100+	140	5	4	4	150
Total (rounded)	390	23,000	17,000	10,000	51,000

16.4 The following lay-out seems easier to follow. It has fewer gridlines, single spacing of rows with deliberate gaps, and some rounding.

	1958	1971	1977	1978
Stock of Dwellings (millions)	16.2	19.5	20.9	21.1
% Owner occupied	39	50	54	54
% Rented from Local Auth.	25	31	32	32
Av. Price of Dwelling (£)	3,200	5,600	14,000	16,000
New Dwellings Completed ('000)				
Public	150	170	170	140
Private	130	200	140	150

Changing the rows into columns as below makes the trends still clearer. Long row captions can be dealt with by using several lines and abbreviations (possibly amplified by footnotes, depending on the context).

	Stock of Dwellings			Average Price of Dwelling (£)	New Dw. Completed ('000)	
	Number (millions)	Owner occup.	Rented from L.A.*		Public	Private
1978	21.1	54%	32%	16,000	140	150
'77	20.9	54%	32%	14,000	170	140
'71	19.5	50%	31%	5,600	170	200
'58	16.2	39%	25%	3,200	150	130

* Local Authority

16.5   (i) Figures are easier to compare reading down than across.

(ii) Rows in single spacing.

(iii) Arranging rows (or columns) in order of size.

(iv) Fewer grid lines.

(v) Reducing the size of typed tables (by about 10 percent linearly).

(vi) Rounding.

The principle works because it allows the reader to compare the relevant figures with fewer eye-movements or other visual interruptions, and thus less interference with one's short-term memory.

Chapter 17: Graphs

17.1   Plotting the data as a rough graph shows quickly that whilst height increased fairly steadily over the years (a linear relationship with age), weight increased relatively faster for the older boys (a curve).

To see this from the table is much more laborious. We would have to work out that average height increases by 19 inches from 5 to 9 years and by 17 inches from 9 to 13 years. Thus the rate of increase is *fairly* constant, indicating a straight line. The increase in weight from 5 to 9 years is 20lb, rather less than the 26lb increase from 9 to 13 years. This implies a curved relationship. (These calculations are time-consuming compared with looking at a graph, but they do give us more quantitative detail.)

17.2   (a) By about £1000 million in each case.

(b) By about £5000 million and £4500 million respectively. The figures are tedious to read off from the graph and they are inaccurate, especially for question (a).

17.3   (a) About the same.

(b) Males aged 15–64 exceeded females, probably by about 2 or 3 million.

17.4   The graph may be satisfying to produce and it looks interesting. But it is not easy to get much detailed information from it.

17.5   (a) A simple message.

(b) In both situations one is communicating to people who did not know the results before—the analyst himself when faced with new data and the audience.

Graphs are, therefore, generally useful when communicating to people who are ignorant about the particular topic.

Chapter 18: Words

18.1 Giving all the main findings at the beginning of a report means they are where the reader wants them. The rest of the report can also be kept briefer. The disadvantage of such a report structure is that it usually requires more work from the author.

18.2 (a) Brevity in a technical report means that more people will read it, and that it is easier to read. But it requires work and self-criticism from the author.
  (b) (i) Giving the main findings first.
      (ii) Omitting things the reader does not need.
      (iii) Cutting down on verbiage.
      (iv) Getting the help of a friendly critic.

18.3 'Courses on marketing strategy are now a major growth area, both in the UK and the US. The first real textbook, Abell and Hammond (1979), went out-of-stock within six months.
  Why? Kiechel (1979) has linked the trend in courses to the growth of strategic consultancy, whose tools seem to offer senior executives a way to cope with the complexity of diverse organizations. Approaches like the now infamous market share/growth matrix (Hedley 1977) have been well publicized, so that our students and their employers demand to know them. But in meeting this demand, marketing teachers may fail in two respects. First, we may teach . . .'
  About 100 words, instead of 170. (Note that giving references with date of publication is more informative.)

18.4 Most technical terms are relatively long, but are a useful short-hand if used often enough to make it worth learning them. A fog-factor of 3, which is readable, still allows an average of 3 long words per sentence.

18.5 'Accounting has grown in importance. It started with the running of estates, traders, and small corporations. Now it is a large part of management.
  Developments in accounting relate not only to financial management. They also relate, both in the private and the public sector, to how organizational structures are seen; to how they function; and to how power and influence is reinforced or created.
  What accountants measure can shape people's views of what is important. The economic and organizational concepts that are implied in accounting can help to create a view of what the organization is like. At a broader social level . . .'
  A fog-factor of $2\frac{1}{2}$ instead of 13. Nuances have been lost (e.g. 'embryonic corporation'); but they would hardly have been noticed by the newcomer to the subject whilst already being known to the knowledgeable.

18.6 (a) How much your colleagues already know.
      How much discussion there will be and what form it will take.
      What questions will be asked.
      What hand-outs should be given and when: at the time, beforehand, or afterwards.
      Which kind of visual aids are possible.
      What action is likely to occur (further work, decisions, etc.)

What you are hoping for.
How long you have (or want) to speak.
What you want to say.
How you will organize it.
How you will rehearse it.

(b) What you want to say.
What you can or should leave out.
Whether a written paper is needed or desirable.
When any hand-out has to be issued.
Which visual aids are possible or desirable.
When you can check them.
When you can check the actual room and have a final run-through.

Chapter 19: Description

19.1   We would need 'norms' for repeat-viewing. For example, what is repeat-viewing like for the 7 pm News on Channel X on other pairs of nights; the Channel X News at other times; for the News on other channels; and for other types of programs. (In practice, it is found that repeat-viewing is generally between 45 and 65 percent, irrespective of program type, so that the initial 55 percent result is 'normal'.)

19.2   The law is an empirical generalization which has been found to hold under a wide range of circumstances when the substance through which the light passes is uniform, but not when it is not uniform. The exceptions cited are therefore not part of the conditions under which the law holds, but are themselves regular.

19.3   To say that one is repeating a measurement means that certain aspects have to be the same (i.e. they have to be 'controlled'): the same apparatus (or kind of apparatus), certain technical procedures (e.g. in measuring people's heights, making sure that they stand up straight). But there are other conditions which are deliberately varied, or allowed to vary, to try to increase the range of generalization (e.g. whether it was done in the morning or the afternoon).

19.4   Correlation does not mean causation. There will nonetheless be causal connections, but not necessarily just 'More pay, therefore more sales.' Thus if sales go up and the chief executive does *not* get more pay, he is likely to leave.

19.5   No. The relationships between the variables which occur explicitly in scientific laws are usually descriptive: saying that this is roughly how x varies when y varies under such and such conditions (e.g. distance travelled and time taken). But there may be underlying causal mechanisms.

Chapter 20: Explanation

20.1   The statement tells us that rust contains iron and oxygen but does not explain the mechanism (e.g. that water is involved, that the process is electro-chemical, etc.). However, the statement does rule out some other possible explanations, e.g. that rusting is simply a change within iron as such, that substances other than oxygen are involved in a major way, and so on.

20.2 The example illustrates that the causal connections between variables like advertising and sales are not simple:
- Expected sales appear to have caused changes in the advertising budgets.
- The correlation appears negative: the more advertising the lower the sales in 1980, and vice-versa in 1981. But the effects *may* still be positive: with less advertising in 1980, sales might have been lower still.
- Instead of advertising generally causing sales, a contrary view is that much advertising is defensive—to safe-guard the existing sales level, rather than to increase it.

20.3 No. Economic theory implies 'other things being equal'. In the example, Brand A may be of better quality, have better packaging, better after-sales service, or more promotional support. Consumers tend to regard a higher price as an indication of quality (and usually they are right).

20.4   (i) Did the R & D department compare their estimate of manufacturing costs with competitors' retail prices or ex-factory prices, or with competitors' actual manufacturing costs? (How would they know the latter?)
  (ii) Did the estimate take into account marketing costs like packaging, storage and transport, selling and promotion, bad debts, replacing faulty product, etc.?
 (iii) How much do the R & D people know about the large-scale manufacturing costs of a product that is new to the firm?
 (iv) Did the estimate include costs of overheads, of capital, of the R & D work itself, etc.?
  (v) Did the estimate allow for a profit margin?

Chapter 21: Observation and Experimentation

21.1 The results were probably induced by the selection effect called regression towards the mean. Checks on this would be to see
- Whether children with high IQs have parents of lower IQs.
- Whether parents with below-average IQs tend to produce brighter children than themselves.
- Whether there is any overall decline generation by generation in IQ levels (or in other 'measures of intelligence').

21.2 Salesmen with high 1979 sales were much more likely to exceed their 1980 targets than those with low 1979 sales. Allowing for their 1979 sales, slightly more salesmen on salary exceeded their 1980 targets than those on commission. (This does not necessarily imply improved selling performance; it may reflect the way the 1980 sales targets were set.)

    The overall figures say that more salesmen on commission exceeded their 1980 sales targets (44 percent) than those on salary (33 percent). This is accounted for by the much larger proportion of salesmen on commission who had high 1979 sales (80 percent compared with only 10 percent of those on salary).

21.3 Physicists already have a backlog of generalizable knowledge about temperatures, i.e. that under certain specifiable conditions water generally boils at 100 °C and freezes at 0 °C. (They check only occasionally that their instruments have not gone out of control, or whether something unexpected has happened to their conditions of observation.)

21.4 The main conclusions are:

  (i) Giving the drug went with increased recovery rates. At low dosage the rate was almost three times higher than without treatment (up to 60 percent versus 20 percent).
 (ii) The high dosage led to a lower recovery rate. (Was this due to side-effects? The randomized allocations of patients makes it unlikely that those given the higher dosage had more severe symptoms or less chance of recovery.)
(iii) The higher level of nursing care went with slightly higher recovery levels for people given the drug. It had a very marked effect for people given the 'dummy' placebo, but no effect for patients given no apparent drug treatment. At the normal level of nursing care, the placebo had little if any effect.

The experimental design has eliminated many alternative interpretations of the data (as illustrated in (ii)). But many conclusions are still unclear, such as those for high dosage. This inconclusiveness would become worse if another similarly well-designed experiment gave rather different results, like no improvement due to treatment. There would then be factors affecting recovery that had not yet been allowed for or understood.

Index

Basic Formulae

(Listed in order of discussion in the text.)

Mean $m = \dfrac{\text{Sum}(x)}{n}$.

Weighted mean of sets A and B $= \dfrac{\bar{x}_A n_A + \bar{x}_B n_B}{(n_A + n_B)}$.

Mean of a frequency distribution $= \dfrac{\text{Sum}(x \times f)}{n}$, where f readings take the value x.

Mean deviation md $= \dfrac{\text{Sum}|x - \bar{x}|}{n}$.

Var $(x) = \dfrac{\text{Sum}(x - \bar{x})^2}{(n - 1)}$ or $\dfrac{\text{Sum}(x^2) - n(\bar{x})^2}{(n - 1)}$.

Variance of a frequency distribution $= \dfrac{\text{Sum}[(x - \bar{x})^2 f]}{(n - 1)}$.

Standard deviation $s(x) = \sqrt{\text{var}(x)}$.

Range $= x_n - x_1$.

Coefficient of variation CV $= \dfrac{s}{\bar{x}} \times 100\%$.

Standardized form of variable $x = \dfrac{x - \bar{x}}{s}$.

Factorial of the number n is $n! = n(n - 1)(n - 2) \ldots (3)(2)(1)$.

Mean of a Binomial distribution $= np$.

Variance of a Binomial distribution $= npq$.

$q = (1 - p)$.

Standard error of the mean $= \dfrac{\sigma}{\sqrt{n}}$.

Estimated standard error of the mean $= \dfrac{s}{\sqrt{n}}$.

t-ratio $= \dfrac{m - \mu}{(s/\sqrt{n})}$.

Correlation coefficient $r = \dfrac{\text{cov}(x, y)}{s_x s_y}$.

$\text{Cov}(x, y) = \dfrac{\text{Sum}(x - \bar{x})(y - \bar{y})}{(n - 1)}$ or $\dfrac{\text{Sum}(xy) - n\bar{x}\bar{y}}{(n - 1)}$.

Standard error of $r = 1/\sqrt{(n - 2)}$ if $\rho = 0$.

Equation of a straight line: $y = a + bx$, where a and b are either any numbers or the values given by regression analysis, depending on the context.

Intercept-coefficient $a = \bar{y} - b\bar{x}$.

Slope-coefficient $b = \dfrac{y_j - y_i}{x_j - x_i}$ for any two points on the line.

Residual standard deviation rsd $= \sqrt{[\text{Sum}(\text{residuals})^2/(n - 1)]}$.

Variance of residuals $= s_y^2(1 - r^2)$.

rsd $= s_y \sqrt{(1 - r^2)}$.

Regression coefficient $b = \text{cov}(x, y)/\text{var}(x)$ for regression of y on x.

Standard error of regression coefficient $b = \dfrac{\text{rsd}}{\sqrt{[\text{Sum}(x - \bar{x})^2]}}$

Regression coefficient $b = \dfrac{\text{cov}(x, y)}{\text{var}(y)}$ for regression of x on y.

Variance-ratio $F = \dfrac{\text{Variance estimate based on sample means}}{\text{Variance estimate based on individual readings}}$.

Standard error of proportion $p = \sqrt{[p(1 - p)n]}$ or $\sqrt{(pq/n)}$.

χ^2-statistic $= (\text{Observed} - \text{Expected Frequency})^2/\text{Expected Frequency}$.

Basic Symbols

a	Intercept-coefficient of a straight line.
α	(alpha) population value of intercept-coefficient.
b	Slope-coefficient of any straight line, *or* that of a regression equation.
β	(beta) population value of slope-coefficient.
CV	Coefficient of variation.
cov (x, y)	Covariance of x and y.
e	Mathematical constant $= 2.718$.
f	Frequency.
m	Mean of readings or sample mean.
m_x	Mean of x readings.
μ	(mu) mean of population.
md	Mean deviation.
n	Number of readings in a set or sample (or number of pairs of readings x, y for two variables).
n	Fixed size of each set in a Binomial distribution.
N	Number of items or events in the population sampled.
p_A	Probability of A.
p_r	Proportion of readings with the value r.
p	Overall proportion of items or events having a specified characteristic.
π	(pi) population value of the proportion.
q	Overall proportion of items or events *not* having the specified characteristic $(= 1 - p)$.
r	Correlation coefficient.
R	Multiple correlation coefficient.
ρ	(rho) hypothesized population correlation coefficient.
rsd	Residual standard deviation.
s	Standard deviation of a set or sample of readings.
sd	Standard deviation.
$s(x)$	Standard deviation of the x readings.
s^2	Variance.
$s/\sqrt{n}$	Estimated standard error of the mean.
$\sigma/\sqrt{n}$	Standard error of the mean.
σ	(sigma) standard deviation of population.
Σ	(Sigma) Summation.
t	Student's t-ratio.
var (x)	Variance of the x-readings.
x	A variable which can either represent *any* value or some particular value, depending on the context.

$\bar{x}$	Mean of the x readings.
$\lvert x \rvert$	Absolute value of the variable x.
$(x - \bar{x})$	Deviation of x from the mean.
y	A variable.
$!$	Factorial
$\doteq$	Approximately equal.